Electrical and Electronic Applications 2

Electrical and Electronic Applications 2

David W. Tyler, CEng, MIEE

Senior Lecturer in Electrical Engineering
at Reading College of Technology

NEWNES–BUTTERWORTHS
TEC
TECHNICIAN SERIES

THE BUTTERWORTH GROUP

UNITED KINGDOM
Butterworth & Co. (Publishers) Ltd
London: 88 Kingsway, WC2B 6AB

AUSTRALIA
Butterworths Pty Ltd
Sydney: 586 Pacific Highway, Chatswood, NSW 2067
Also at Melbourne, Brisbane, Adelaide and Perth

CANADA
Butterworth & Co. (Canada) Ltd
Toronto: 2265 Midland Avenue, Scarborough, Ontario M1P 4S1

NEW ZEALAND
Butterworths of New Zealand Ltd
Wellington: T & W Young Building, 77–85 Customhouse Quay, 1,
    CPO Box 472

SOUTH AFRICA
Butterworth & Co. (South Africa) (Pty) Ltd
Durban: 152–154 Gale Street

USA
Butterworth (Publishers) Inc
Boston: 10 Tower Office Park, Woburn, Mass. 01801

First published 1980

British Library Cataloguing in Publication Data

Tyler, David W
    Electrical and electronic applications 2.

    1. Electric engineering
    I. Title
    621.3   TK145   79–41058

    ISBN 0–408–00412–6

Typeset and produced by Reproduction Drawings Ltd, Sutton, Surrey
Printed in England by Page Bros Ltd., Norwich, Norfolk

Preface

This book covers the syllabus of the T.E.C. Unit U76/361, 'Electrical and Electronic Applications 2'. The applications require a knowledge of the principles involved and these are covered briefly in the text. However, if further principles examples and discussion are required, a parallel volume 'Electrical Principles for Technicians 2' by S. A. Knight may be referred to. S.I. units are used throughout.

There are two groups of problem examples included in the text of most sections throughout the book. One group comprises examples, worked out for you, which illustrate method and procedure and these are prefixed by the heading 'Example'. The second group are self-assessment problems which are for you to work out before proceeding to the next part of the text; these problems are simply given a number in parentheses ( ). They are intended to illustrate those parts of the text which immediately precede them although occasionally they refer to earlier work as well. All examples of either group are numbered through in order so that the answers to numerical questions may be found at the end of the book without difficulty. At the end of each section there is a further selection of problems. In a subject such as this, which involves largely descriptive work, many of the questions are of necessity non-numerical and answers to these must be sought in the text.

I wish to thank the Institution of Electrical Engineers for permission to quote from *Regulations for the Electrical Equipment of Buildings*. Any interpretation of those Regulations is mine alone.

D. W. Tyler

Contents

1 Transmission and distribution of electrical energy

Aims: At the end of this section you should be able to:
Explain why transmission is carried out at very high voltages.
Understand the factors which affect the design and arrangement of the transmission and distribution system.
Compare overhead lines with underground cables.
Explain the purpose of switchgear.
Describe the equipment and layout of a small distribution substation.

SYNCHRONOUS GENERATORS

Virtually all the generation of electrical energy throughout the world is done using three-phase synchronous generators. Almost invariably the synchronous generator has its magnetic field produced electrically by passing direct current through a winding on an iron core which rotates between the three windings or phases of the machine. These windings are embedded in slots in an iron stator and one end of each winding is connected to a common point and earthed. The output from the generator is taken from the other three ends of the windings. The output from a three-phase generator is therefore carried on three wires. In many three-phase diagrams single line representation is used when each line on the diagram represents three identical conductors. *Figure 1.1* is drawn using this method.

All such generators connected to a single system must rotate at exactly the same speed, hence the term synchronous generator.

They are driven by prime movers using steam generated by burning coal or oil or by nuclear reactors, water falling from a higher to a lower level, or aircraft gas turbines burning oil or gas. A very small amount of generation is carried out using diesel engines.

Generators range in size from 70 MVA (60 MW at 0.85 power factor) at a line voltage of 11 kV which were mostly installed in the 1950s, through the intermediate size of 235 MVA (200 MW at 0.85 power factor), to the recent machines rated at 600 MVA (500 MW) which generate at 25.6 kV. There are generators rated at 660 and 1000 MW but these are rare at the moment.

ECONOMICS OF GENERATION AND TRANSMISSION

The power in a single phase circuit $= VI \cos \phi$ watts where V and I are the r.m.s. values of circuit voltage and current respectively and ϕ is the phase angle between the current and voltage.

As an example, consider a power of 1 MW at 240 V and a power factor of 0.8 lagging. (1 MW = 1000 kW = 10^6 watts.)

$$240 \times I \times 0.8 = 10^6$$

$$I = \frac{10^6}{240 \times 0.8} = 5\,208 \text{ A}$$

By increasing the voltage to say 20 000 V the required current falls to 62.5 A.

The voltage drop in a transmission line due to the resistance of the line
= IR volts.

The power loss = voltage drop × current flowing

$$= IR \times I$$

$$= I^2 R \text{ watts.}$$

Using the above values of current it may be deduced that:

1.   for a conductor of given size and resistance, the line losses at 240 V and 5208 A will be very much greater than at the higher voltage; or

2.   if the losses are to be the same in both cases the conductor for use at 240 V will need to have a very much lower resistance and hence have a much larger cross-sectional area than that for use at the higher voltage.

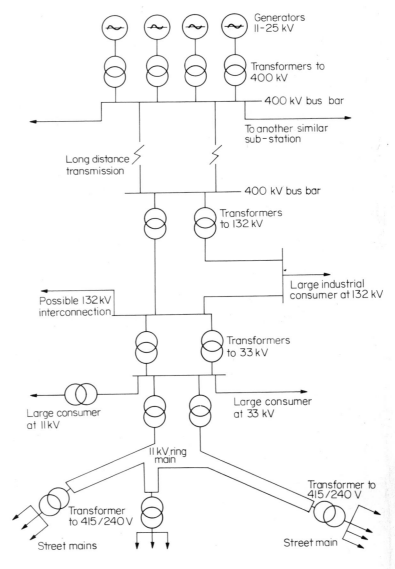

Figure 1.1

To enable large powers to be transmitted through small conductors while keeping the losses small therefore requires the use of very high transmission voltages. The voltages commonly in use in the UK are shown in *Figure 1.1*. At each stage circuit breakers would be employed but these are not shown for simplicity.

Each generator feeds directly into a step-up transformer which increases the voltage to 400 kV. Power is transmitted to the major load centres at this voltage where it is transformed down to 132 kV (sometimes an intermediate step at 275 kV is used). Some heavy industry is fed at this voltage but most of the 132 kV system forms local distribution to 33 kV substations. These feed industry and a series of 11 kV substations. Ring mains at 11 kV feed transformers which supply power at 415/240 V to domestic and commercial consumers.

To save money on transformers it would seem that generation could best be carried out at 400 kV but so far it has not been found possible to develop insulation for use in rotating machines which will withstand such a high voltage while allowing the heat produced in the winding to be dissipated. In particular there are problems where the conductors leave the slots in the iron core and emerge into the gas filled spaces at the ends of the stator. The problems are overcome in transformers for use at 400 kV by the use of paper insulation and immersing the windings and core completely in a special oil which insulates electrically and convects heat away.

Cables for extra-high-voltage work are also paper insulated and contain oil under pressure. They are laid individually and heat is conducted away by the soil.

As the voltage is increased, as we have already seen, the size and hence cost of the conductors decreases. However as the voltage is increased the cost of insulation is increased. Cable insulation becomes thicker, oil is used and this must often be maintained under pressure which requires additional plant. Very expensive cable terminations called *sealing ends* have to be used.

Switchgear for use at high voltages is more complicated, bulkier and more expensive than that for use at medium and low voltages. When a circuit breaker opens to interrupt a circuit an arc is drawn between the contacts. At domestic voltages the arc is small and arc extinction occurs quickly in the atmosphere. At extra-high voltages the arc is much more difficult to extinguish and air or oil often under pressure have to be used. In addition, all the electrical parts must be kept well away from earth and these clearances are much greater where very high voltages are used.

The capital costs of extra-high-voltage gear reflect voltage levels but are not affected very much by the cross-sectional area of the conductors used.

Figure 1.2 shows comparative costs of conductors and insulation for increasing system voltages.

In addition to the cost of equipment there is the provision of land to consider. The additional bulk of extra-high-voltage gear means that substations may occupy land areas of hundreds or even thousands of square metres.

In *Figure 1.1* we see that major transmission is at 400 kV. Since the power transmitted from a single power station may be in excess of 2 000 MW, the high cost of insulation and switchgear is justified by the considerable reduction in conductor costs. At distribution level the supply for a single factory or for housing from an individual transformer represents a relatively small power so that even at much lower voltages

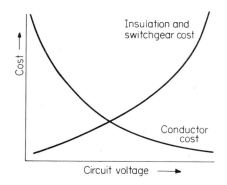

Figure 1.2

the current involved is quite small. The cost of extra-high-voltage switch-gear would not be justified and the land area for a substation might well be restricted.

As local demands decrease the voltage at which they are supplied is reduced. A large factory requiring 100 MW will be fed directly from either the 132 kV or 33 kV system. A smaller factory requiring only 1 MW could be fed from the 11 kV system whilst a group of houses and shops with a collective requirement of 500 kW will be fed at 415/240 V. The conductor cross-sectional areas generally lie between 225 mm² and 650 mm² irrespective of voltage, the insulation and switchgear costs and the land area per substation decreasing at each successive voltage reduction.

Typical transformer ratings at the various voltage levels are:

25.6/400 kV	600 MVA
400/132 kV	150–250 MVA
132/33 kV	50–75 MVA
33/11 kV	10-15 MVA
11 kV/415/240 V	250-500 kVA

OVERHEAD LINES

Overhead lines for power transmission are almost invariably made of aluminium with a steel core for strength. The bare conductors are supported on insulators made of porcelain or glass which are fixed to wooden poles or steel lattice towers.

Figure 1.3 shows some typical British line supports together with the associated insulators. All the steel lattice towers shown use suspension insulators whilst the wooden poles may use either type. Three conductors comprise a single circuit of a three-phase system so that the 33 kV single circuit tower has three cross arms and three suspension insulators. Towers with six cross arms carry two separate circuits.

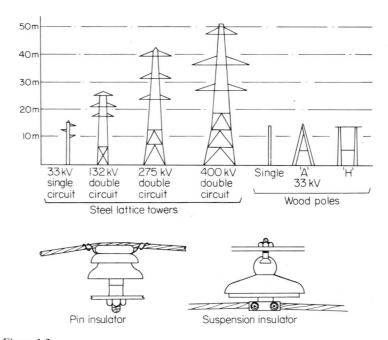

Figure 1.3

On high-voltage lines each support must carry a consecutive recognition number and a circuit identifying colour disc. The supports must be capable of supporting the line without movement in the ground when both line and supports are carrying a specified ice loading and an 80 km/hour wind is blowing. Safety factors of 2.5 for steel towers and 3.5 for wood must be allowed. A safety factor of 2.5 means that with the ice and wind loading the load is 1/2.5 of that which would cause the support to collapse.

Wood supports are red fir impregnated with creosote and may be in the form of single poles or two poles made into an A or H. In the UK they are used for circuits up to 33 kV but in other countries lines up to 250 kV using 50 m poles have been erected. Since Britain imports most of the trees required and each pole is in fact a complete tree trunk, large ground clearances using poles proves to be extremely expensive.

Towers are made of steel angle section and may easily be fabricated up to almost any height by adding extra bottom sections or trestles.

COMPARISON BETWEEN OVERHEAD LINES AND UNDERGROUND CABLES

Cost The overhead line is air insulated and is supported on insulators mounted on towers or poles which are 100–400 m apart. The underground cable is fully insulated and armoured to protect it against mechanical damage and then covered overall with a corrosion resistant material.

For extra-high voltage work the overhead line is made of steel cored aluminium while the underground cable is made of copper to reduce

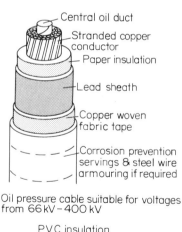

Oil pressure cable suitable for voltages from 66 kV – 400 kV

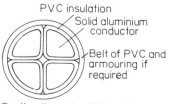

Section through solid aluminium conductor cable for use as street mains (415/240 V)

Figure 1.4

the resistance of a given cross-sectional-area cable. Local underground distribution cables may use solid aluminium cores which are insulated with p.v.c. Four such cores are laid up, armoured and served overall to form a three-phase cable with the fourth conductor as neutral or earth connection. This is shown in *Figure 1.4* together with an oil pressure cable, three of which are required to form a three-phase circuit.

The high cost of copper, insulation, armouring and corrosion protection, together with that of taking out a suitable trench and refilling it, make the e.h.v. underground cable many times more expensive than the overhead line. The price difference at 415/240 V using aluminium cables is not so great and these may be preferred on environmental grounds.

Environment   The underground cable is invisible. However there can be no building over it or large trees planted since in the event of a fault it must be possible to dig a suitable hole to effect a repair. The heat produced by an e.h.v. cable can effect the soil around it thus modifying the plant growth in the immediate vicinity.

The overhead line has conductors and supports which are sometimes visible for long distances. Electrical discharges from the lines can cause radio interference.

Reliability   There is little difference in the reliabilities of the two systems. The overhead line can be struck by lightning whereas the underground cable is at the mercy of earth moving machinery especially when roads are re-made or trenches for other services are dug. Occasionally a cable will develop a small hole due to movement over a stone for example giving rise to water ingress followed by an explosion but this is thankfully rare.

Fault finding   Overhead lines are patrolled regularly on foot or by helicopter. Broken insulators can be seen and by using infra-red detection equipment local hot spots can be found possibly in compression joints where two lengths of conductor have been joined. Repairs are reasonably cheap since the line can be taken down, insulators replaced, joints remade and towers repainted at almost any time. If an underground cable develops a fault electrical methods have to be used to locate it. Unless the route is precisely known and the test accurately carried out a great deal of digging is required before the fault can be found. When it has been located the repair is expensive especially on e.h.v. cables.

SWITCHGEAR, DEFINITIONS AND USES

Circuit breaker.   A circuit breaker is a mechanical device for making and breaking an electrical circuit under *all* conditions.
Switch.   A switch is a device for making and breaking a circuit which is carrying a current not greatly in excess of normal loading.
Isolator.   An isolator is a means of isolating or making dead a circuit which is not carrying current at the time (like pulling out a fuse in the home so that work may be carried out in safety on a circuit). The isolator may be used to close a circuit on to load.

The Electricity Supply Regulations state that no piece of electrical equipment may be connected to the mains unless the circuit incorporates a device which will disconnect that equipment automatically in the event of a fault.

According to the definitions above, a circuit breaker is such a device. These are made in miniature form for domestic use with current ratings of between 5 A and 60 A at 240 V while there are larger sizes for

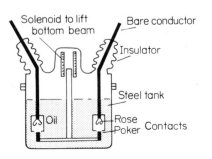

Figure 1.5 Bulk oil circuit breaker

industry and transmission and distribution substations which can deal with the highest voltages and currents presently in use.

A fuse is often used in place of a circuit breaker in circuits operating up to 11 kV but once it has operated to clear a fault it has to be replaced. This takes time and the larger sizes are very expensive. A circuit breaker can be reclosed after clearing a fault and in addition it may be useful in the rôle of a switch, making and breaking circuits under normal conditions.

Switchgear may be of either the indoor or outdoor variety. For use indoors all electrical conductors are completely enclosed. For use outdoors the circuit breakers are made completely weatherproof. The circuit conductors and isolators are enclosed in 11 kV to 415/240 V substations but at higher voltages they are bare metal insulated from earth using porcelain or glass insulators as described for overhead lines.

Figure 1.5 shows a bulk oil circuit breaker for use outdoors. It is suitable for use up to 132 kV. *Figure 1.6* shows an air circuit breaker employed up to about 11 kV. It may be used indoors or outdoors according to the type of enclosure.

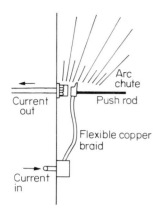

Figure 1.6 Air circuit breaker

SUBSTATION LAYOUT AND EQUIPMENT

The small substation shown in *Figure 1.7* is typical of the thousands of such installations feeding factories, housing and commercial premises. It is connected to the 11 kV ring main shown in *Figure 1.1*.

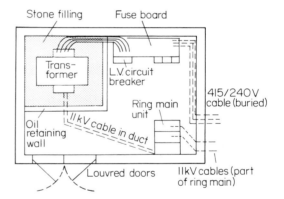

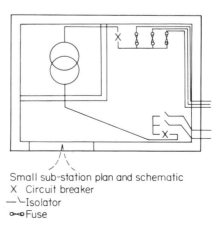

Small sub-station plan and schematic
X Circuit breaker
—⌐Isolator
o—∞Fuse

Figure 1.7

Considering *Figure 1.1*; the ring main is equipped with two isolators at each tap off point and these are normally closed making the 11 kV ring complete. The transformer is fed through a circuit breaker. If a fault occurs on the transformer itself the circuit breaker should open. If a fault occurs on the low voltage system it should be cleared by the low voltage circuit breaker or one of the fuses. Should a fault occur and the correct clearance not take place the circuit breakers controlling the 11 kV ring main (not shown) would operate so making the whole ring dead, depriving other consumers of their supply.

Circuit breakers may be of the oil immersed or air break type. Switches would be used to control individual low voltage circuits for lighting and heating and an automatic switch called a contactor would be used for motor control.

The transformer in the substation will be rated at 11 000 to 433 V nominally. On the high voltage side of the transformer there will be a tap changer. This is a manually operated device which is adjusted off load to alter the number of turns on the winding and has the effect of altering the output voltage to the desired level. The transformer rating will be 300, 500, 750 or 1000 kW. It will generally be oil filled using special transformer mineral oil and the heat produced will be carried by the oil to cooling tubes on the outside of the transformer whence it is convected away by natural air circulation. For this reason the building must be adequately ventilated. This is achieved by fitting louvres in the door of the substation and air bricks or more louvres high up in the walls. Very large transformers are fan cooled but these are installed outdoors. The transformer is either mounted on a plinth in the centre of a large pit filled with stones or there is a small wall built around it. If the transformer suffers a fault and oil leaks out it must be contained to minimise the risk of fire spreading.

In a very large substation with many circuit breakers fire resistant barriers are used to create small sections and automatic fire fighting equipment is fitted in each section using either carbon dioxide gas or fine water spray. Should a fire occur on one circuit breaker only the equipment in the affected section operates.

PROBLEMS FOR SECTION 1

(1) Calculate the value of current necessary to transmit a power of 100 kW at a power factor of 0.7 lag at (i) 500 V (i) 5000 V. If in each case the conductor resistance is 0.2 Ω calculate the two power losses.

(2) What is the function of the 400 kV system in the UK?

(3) What is the function of the 132 kV and 33 kV systems?

(4) What type of insulation is used in e.h.v. underground cables?

(5) Why does the cost of transmission equipment rise as the voltage levels are raised?

(6) In the UK domestic premises are supplied at between 230 and 250 V. In some other countries a 110 V system is used. What advantages and disadvantages are there to such a system?

(7) What advantages are there to using steel lattice towers as compared with wood poles?

(8) List the advantages and disadvantages of an overhead transmission system as compared with an underground cable system.

(9) What is the function of a circuit breaker?

(10) What advantages has a miniature circuit breaker over a fuse?

(11) What difference in design would you expect to find in an outdoor circuit breaker as compared with one situated indoors?

(12) What is the function of the stone filled pit beneath an oil filled transformer?

(13) What is the function of a tap changer?

(14) Why are large indoor switching stations sectioned?

2 L.V. distribution systems, reasons for earthing and simple circuit protection

Aims: At the end of this section you should be able to:
Describe typical three-phase industrial installations.
Explain why earthing is carried out.
Describe voltage and current differential earth leakage protection.
Describe the construction and operation of H.B.C. fuses and the situations in which they are used.
Describe various overcurrent relays.
Calculate the current distribution in, and the efficiency of, radial feeders and ring mains.

THE THREE-PHASE SYSTEM

In the three-phase generator the magnetic field on its rotor links in sequence with three equally spaced windings or phases on its stator inducing sinusoidal voltages with equal maximum values in each of them.

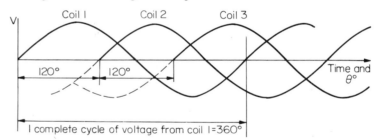

Figure 2.1

These are often known as the Red, Yellow and Blue voltages respectively. These are shown in *Figures 2.1* and *2.2*. The phasors rotate anticlockwise and the red phase voltage wave is followed by that of the yellow phase and again by that of the blue phase.

Figure 2.3 shows a schematic diagram of a generator, its windings physically displaced from each other by 120° and with one end of each winding connected to earth. Because of its appearance this is known as the star connection. The supply lines are labelled red, yellow and blue respectively and the wire connected to the common point is known as the neutral. The voltage from any output line to the neutral is called the phase voltage and the voltage between any pair of supply lines is called the line voltage.

The line voltage = $\sqrt{3}$ × the phase voltage

For example where the phase voltage is 240 V, the line voltage is $\sqrt{3}$ × 240 = 415 V. Single phase loads are connected between any line and the neutral wire and this is the normal situation in the home. One house in a street is connected between the red line and neutral, the next

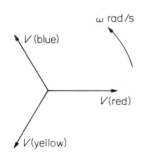

Figure 2.2 Phase voltage phasors in a 3-phase system

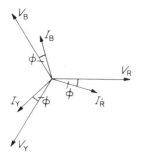

Figure 2.4

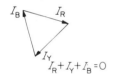

Single phase loads   **Figure 2.3**

house between the yellow line and neutral and the next from blue to neutral.

In industry three-phase loads may use three or four wires.

Load balancing   Consider now the case of three single-phase loads being equal in magnitude and phase.

Figure 2.4 shows the three phase voltages as in *Figure 2.2* together with the current phasors. These are all equal in magnitude and lag on the phase voltages by an angle $\phi°$. All these currents flow from their respective lines to the neutral. The current in the neutral wire must therefore be the sum of these three currents.

Figure 2.5 shows the phasor addition so that the neutral current proves to be zero. There is no current flowing in the neutral wire back to the supply generator when the loads are balanced.

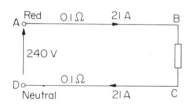

Figure 2.5

It is desirable that over the country as a whole the loads shall be balanced over the three phases since this minimises the cable losses and voltage drops in the supply lines. In addition, there are problems in the generators themselves when the three phases are not equally loaded.

Let us consider the power loss and voltage drops in an unbalanced system.

The worst out of balance is where all the load is on a single phase.

Figure 2.6 shows a single phase resistive load drawing a current of 21 A from a 240 V supply along a cable in which line and neutral resistances are both 0.1 Ω.

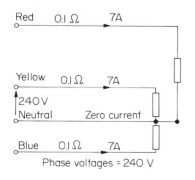

Figure 2.6

The power loss = I^2R watts

In the line, power loss = $21^2 \times 0.1 = 44.1$ W. This is repeated in the neutral wire so that the total loss = 88.2 W.

The fall of potential from A to B = $IR = 21 \times 0.1 = 2.1$ V
The fall of potential from C to D = $IR = 21 \times 0.1 = 2.1$ V

The potential difference from B to C = $240 - 2.1 - 2.1 = 235.8$ V

Now consider the effect of obtaining the same power (closely) by using three separate single-phase resistive loads each drawing 7 A spread out over the three phases. The phase voltages are 240 V as before. There is no neutral current since the loads are balanced.

The power loss in each line = $I^2R = 7^2 \times 0.1 = 4.9$ W
In three lines the power loss = $3 \times 4.9 = 14.7$ W
The fall of potential along each line = $IR = 7 \times 0.1 = 0.7$ V

The potential difference across each load = $240 - 0.7 = 239.3$ V
Balancing the loads over the three phases causes the loss in the neutral wire to be eliminated whilst reducing the voltage drops in the cable.

Phase voltages = 240 V

Figure 2.7

INDUSTRIAL
INSTALLATIONS

General arrangement A typical substation for a factory is shown in *Figure 1.7*. Such a sub-station would most likely be situated on the factory premises. Metering of energy consumed and the maximum demand made by the factory on the supply system is carried out using current transformers fitted in, or adjacent to, the main low voltage circuit breaker or fuses. The current transformers feed a kilowatt-hour meter which at 415/240 V derives its voltage coil supply directly from the bus bars without the use of poten-tial transformers (see Chapter 6 for the use of current and potential transformers). The main distribution fuse board is fitted in the substation.

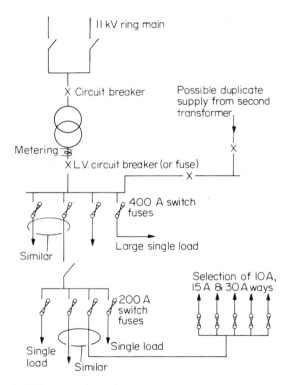

Figure 2.8 Small factory schematic

In *Figure 2.8* this is shown as being equipped with switches incor-porating fuses rated at 400 A. The actual size employed will depend on the rating of the equipment installed in the factory.

Some large single-phase loads may be fed directly from this point, the actual control being by local switch or contactor. There will be several ways feeding further fuse boards in particular workshops. These will feed local heavy loads or further fuse boards for lighting, heating, small tools and processes.

Types of cable Cables which connect the main substation switchboard to local distribu-tion boards may either be paper insulated and lead covered or PVC insulated and sheathed overall.

Distribution fuse boards

Distribution fuse boards will normally be situated in metal cases fitted with bus bars and will be equipped with fuses to British Standard 88. Every fuse board must be controlled by either the main switchgear or by a larger fuse on a previous fuse board. The arrangement is shown in *Figure 2.8*. Every subcircuit must be connected to a separate way of a distribution fuse board.

Motor control

Where motors are to be controlled contactors will be used into which there will be built some form of overcurrent prevention device. However in the event of a short circuit fault on the motor or the supply cable the fuse should clear the fault in less time than it takes for the contactor to operate.

Wiring systems

(a) Screwed metal conduit. This is steel tube in a number of sizes from about 1.5 cm to 7 cm in diameter. The conductors used in the conduit will be of copper with PVC insulation. The conductors are pulled into the conduits once they are installed in the correct positions. This ensures that conductors can be replaced during the life of the installation. Conduits protect the conductors from all but the most severe mechanical hazards. The tube provides good earth continuity but a separate earth wire is often drawn in. The conduits may be run on the surface of a wall or buried under plaster.

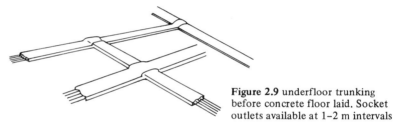

Figure 2.9 underfloor trunking before concrete floor laid. Socket outlets available at 1–2 m intervals

(b) Plastic conduit. This is similar to the metal conduit except that it is made from PVC. It is easier to install being semi-flexible but it does not afford such good mechanical protection. Above about 65°C it tends to soften and it becomes brittle at sub-zero temperatures.

(c) Cable duct systems. The cable duct is like a large conduit but is usually rectangular. It may be up to 30 cm square. It may replace normal skirting board in offices or may be buried under the floor. It can carry services such as the telephone, provide alarm and staff location facilities and allow power supplies to be taken off every metre or so along its length. It is therefore particularly useful in open plan buildings where locations of equipment can change from day to day.

(d) Mineral insulated copper sheathed cable (MICS). The conductors are of solid copper and are held rigidly within a copper tube using very highly compressed magnesium oxide insulation. The main advantage of this cable lies in its resistance to all forms of ill treatment. It can be hammered flat, heated to red heat, immersed in water and many chemicals and it will still continue to function.

(e) PVC insulated cable. This may be stranded copper or solid aluminium made up as a single core, twin cores or twin cores with an earth wire.

(f) Overhead bus bar trunking. This is a very versatile method of distributing supplies in a workshop. A tap-off point is provided every few

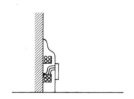

Figure 2.10 Skirting trunking

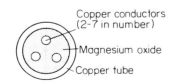

Copper conductors (2–7 in number)

Magnesium oxide

Copper tube

Figure 2.11 M.I.C.C. cable

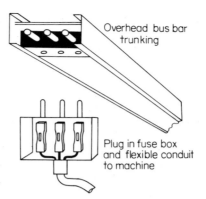

Overhead bus bar trunking

Plug in fuse box and flexible conduit to machine

Figure 2.12

metres and a special fuse box with flexible conduit connections may be plugged in providing supplies for machines and other equipment. The trunking can also carry lighting fittings and this is a very convenient way to provide fluorescent lighting over a working area.

CIRCUIT PROTECTION

Fault level   When considering circuit protection the fault level at the point to be protected must be taken into account. The fault level is the number of volt-amperes which would flow into a short circuit if one occurred at the particular point. On a 415 V, three-phase street main this could be around 2000 kVA which represents a line current of nearly 3000 A. The fault level is limited by the impedance of all the apparatus between the point of the fault and the supply generator. Hopefully, when such a current flowed a protective device such as a fuse would operate so cutting off the supply. At a similar voltage within a factory where the distance from the supply transformer may be less and the transformer itself larger, the fault level could rise to 18 000 kVA which represents a current of approximately 25 000 A.

The nearer the fault is to the supply generators, the larger is the fault level. A fault on the 400 kV system can give fault levels in excess of 20 000 MVA.

Earthing   Both the IEE Regulations and Electricity Supply Regulations require one point on a transformer-fed system to be earthed. Such earthing provides a return path for earth current, so facilitating clearance of faulty circuits.

Power transformer secondaries are star connected and are earthed in exactly the same way as the generators (see *Figure 2.3*). A fault usually exists on a piece of equipment because the live conductor or part of the winding of the equipment comes into contact with the metal casing of the equipment. If this casing is not connected to earth then the equipment becomes potentially dangerous. A person standing on the ground touching the casing of the equipment will receive an electric shock as the potential difference between that of the casing and earth causes a current to flow through the body. It only requires a few milliamperes along a route including the heart to cause death. Earthing the equipment allows current to flow through the earth connection back to the supply transformer neutral. If the impedance of the earth loop is small, sufficient current will flow to melt the fuse in the supply line.

A high earth loop impedance is dangerous since the metal case will remain connected to the supply if insufficient current flows to cause fuse clearance.

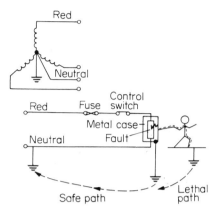

Figure 2.13

In certain situations equipment is run without an earth wire. Such equipment is completely enclosed in a layer of insulating material or so disposed in an insulated box so that it is impossible for the box to become alive or for a person to push a finger in and so touch live metal. A television set in a plastics or wooden cabinet is an example of this type of construction.

Earth leakage protection

In a single phase system the current flowing along the live wire is equal in magnitude to that in the neutral wire unless the circuit has an earth fault. In a three-wire three-phase system the sum of the three line currents in a healthy circuit is zero (*Figures 2.4* and *2.5*). The fuse depends on faults being severe enough to cause very large currents to flow so clearing the fuse. A much better current balance system involves detecting when the currents in the supply lines do not sum to zero and then causing an isolating switch to open. Alternatively the potential difference between the casing and earth can be used to operate an isolating device. When this potential is zero the equipment is healthy.

Protective multiple earthing

This system involves earthing the neutral wire on the distribution network at a number of points instead of just at the supply transformer. This requires special Ministry permission since it contravenes the Electricity Supply Regulations which require one earth point only.

In addition, consumers have their earth and neutral terminals connected together on their premises and these in turn are bonded to the gas and water systems to prevent the potential of these services rising above that of earth. The increased use of plastic main water pipe means that the water service can no longer be regarded as a safe earth in many areas. In the case of the gas service, any rise in potential could cause sparking and an explosion.

The advantages claimed for this system are:

1. Four core cable need not be used. Three core cable with armour-

ing or metallic sheathing which becomes the neutral/earth conductor is cheaper and there is a weight saving.

2. If the neutral/earth conductor becomes discontinuous at any point, say at a joint in the cable, current can continue to flow between the disconnected sections through the ground because of the regular earthing points and the safety and operation of the system is unimpaired.

Earth leakage circuit breakers   *1. Current balance.   Figure 2.14* shows a current balance schematic for a single phase supply. The live and neutral wires pass once or twice round the soft iron core in opposite directions. When the currents in live and neutral wires have the same value the net magnetomotive force in the ring is zero and no flux links the detector coil. When the two currents differ, as they will do if an earth fault is present on the load, then there will be a resultant magnetisation of the ring. By transformer action a voltage will be induced in the detector coil which will give rise to a current in the trip coil and the circuit breaker will open. An out of balance of only a few milliamperes is sufficient to cause operation.

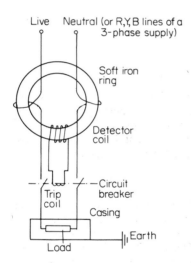

Figure 2.14

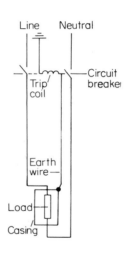

Figure 2.15

2. Voltage operated.   If a fault develops in the load such that the potential of the casing rises above that of earth, a current flows in the earth wire through the trip coil and the circuit breaker opens. Less than 40 V is required for operation.

The H.B.C. fuse   At the fault level in domestic situations a rewireable fuse is quite adequate. In clearing the copper element melts and spreads itself over the surface of the carrier. With very high currents this melting takes place explosively and current may continue to flow in copper vapour. The danger here is that the fault would have to be cleared further back in the network, possibly at the main circuit breaker but when this happens a large part of the system is shut down. (See *Figure 2.8*: if the l.v. circuit breaker opens instead of an individual fuse clearing all the fuse boards become de-energised.)

In this case the circuit breaker has operated as back-up protection for the fuse which should have operated but has failed to do so.

Rewireable fuses are also open to abuse in that they can be rewired using almost any conductor, hairpins and nails for example.

A much better proposition, and indeed essential at high fault levels, is the High Rupturing Capacity or High Breaking Capacity fuse. This comprises a ceramic tube filled with fine sand. It has plated brass end caps of various shapes suitable for clipping or bolting into its fuse carrier. The fusible element is made from silver since this is the best conductor of electricity known. The element may be a single strand or several strands in parallel. A three-strand element is shown in *Figure 2.16*.

Where currents are marginally above that to cause operation the element melts and runs into the sand forming literally thousands of minute breaks in the circuit. Where very large currents are involved the heat produced explosively melts the silver which combines chemically with the sand forming an insulating material and under these circumstances the circuit interruption is extremely rapid taking less than one half cycle of the a.c. supply. The fuse has inverse time characteristics, the larger is the current involved the shorter is the clearance time.

The fuse is non-deteriorating, does not contaminate the fuse carrier when it operates and within fairly close limits the correct fuse must be fitted to a circuit since different ratings have different physical sizes and types of end connections.

The current rating marked on the fuse is that value of current which it can carry continuously without melting or in any way deteriorating.

The rated minimum fusing current is the least value of current which will actually cause the element to melt.

$$\text{Fusing factor} = \frac{\text{Rated minimum fusing current}}{\text{Current rating}}$$

Fuses may be used to provide either close or coarse protection. With close protection the fuse operates when the circuit current is marginally above the current rating whilst with coarse protection a larger overload is permitted.

Cables are rated so that their temperatures do not rise above a safe maximum value above which the insulation deteriorates. This value is $70°C$ for PVC insulated cables for example.

If coarse protection fuses are used there is a possibility of prolonged overcurrent and the cable size has to be increased accordingly. With close protection the possible overload is less. Cable sizes using both types of protection are laid down in the wiring regulations of the Institution of Electrical Engineers. The classes of fuse are:

Class P Fusing factor 1.25 or less. These provide protection for circuits in which virtually no overload is permitted.

Class Q Group 1. Fusing factor between 1.25 and 1.5.
Group 2. Fusing factor between 1.5 and 1.75.
These provide protection for circuits which can withstand some degree of overloading.

Class R Fusing factor between 1.75 and 2.5. Generally these fuses are used as back-up protection for some other form of protection which should normally operate first. This could be a thermal or magnetic relay working in conjunction with a contactor.

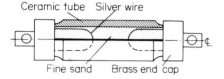

Figure 2.16

Ceramic tube Silver wire

Fine sand Brass end cap

The instantaneous overcurrent relay

In *Figure 2.17* the load current or the output from a current transformer passes through the coil on the iron core causing it to become magnetised. At a certain value of current the magnetic field is strong enough to lift the armature so closing the tripping contacts which will cause the associated circuit breaker to trip. Adjustment is by moving

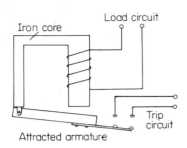

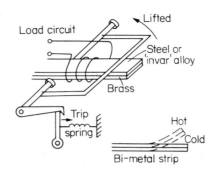

Figure 2.17

Figure 2.18

the armature nearer to or further from the core or by the use of a restraining spring.

The thermal relay
Current flowing either through a heating coil wound on a bimetal strip or through the strip itself causes it to be distorted upwards lifting the cross beams. This occurs since brass expands more than iron or Invar when heated to the same temperature. Lifting the beam releases a trip lever which either mechanically unlatches the associated circuit breaker or closes a pair of tripping contacts as in the instantaneous relay. Adjustment is by moving the starting position of the bimetal strip so varying the amount of distortion necessary to cause tripping.

The magnetic relay
Current flowing in the coil of the relay shown in *Figure 2.19* attracts the iron plunger upwards against the force of gravity. When a pre-determined value of current is reached tripping occurs by lifting the cross beam as in the thermal relay. The current at which the relay

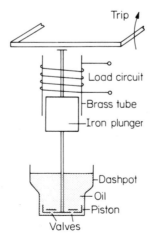

Figure 2.19

operates is adjusted by raising or lowering the dashpot so altering the relative positions of the plunger and coil. For the first few millimetres of travel the piston moves slowly as it is pulled through the oil, the valves being automatically closed by this action. As the piston moves into the wide section of the dashpot the speed increases and the trip bar is knocked upwards. The dashpot prevents instantaneous operation during very short periods of overcurrent. The valves automatically open as the plunger falls allowing the relay to reset rapidly.

The induction overcurrent relay

The output from a current transformer is fed to coil 1 on the laminated E core. This current sets up a magnetic flux in the core, across the gap in which a circular copper disc is situated, and into the bottom U core which is also of laminated construction.

Coils 1 and 2 make up a transformer and the alternating flux set up by the current in coil 1 induces a voltage in coil 2 which is 90° out of phase with the flux. (For a detailed explanation of this see Chapter 6.) This voltage drives a current through the two coils on the U core setting up a flux which crosses the gap into the E core.

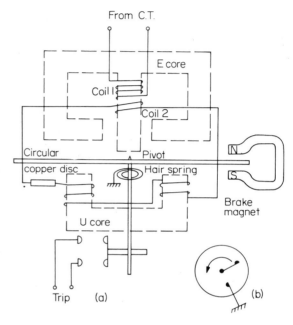

Figure 2.20

There are now two magnetic fluxes linking with the copper disc, both alternating and differing in phase. Under these conditions a torque is set up in the disc which would cause it to rotate if it were free to do so. The device is in fact a small induction motor. The disc is restrained by hair springs and at a predetermined current the torque produced in the disc overcomes that of the springs and the disc turns so closing a tripping circuit (*Figure 2.20b*). The time taken to trip at a particular value of current is adjusted by altering the starting position of the disc so that it has a different distance to travel.

The disc can be made to start turning at a selected value of current by altering the number of turns on coil 1.

Operating characteristics of fuses and relays

Figure 2.21 shows typical characteristics of H.B.C. fuses, induction relays and thermal and magnetic relays. There are many different types designed for specific purposes so that it is only possible to make general comments.

Assuming that the fuse rating and full load currents for the relays are the same and that the induction relay disc makes its full travel, it may be seen from the curves that for currents greater than 2.5 times full load the fuse will clear a fault faster than any of the relays. For currents below this value the induction relay is faster.

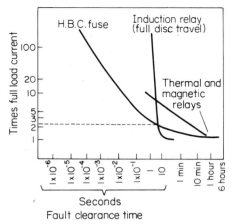

Figure 2.21

There is a similar crossover point between the induction relay and the other relays in the region of 10 times full load current. The induction relay is unable to operate much faster above this value of current due to saturation effects in the relay and the current transformer so that the characteristic is nearly vertical above 10 times full load current.

It follows that where protection is required against short circuit faults the fastest possible clearance time is essential to limit damage due to heating and from the mechanical stresses involved when large currents flow in conductors in close proximity. The H.B.C. fuse gives this fast clearance time. From the characteristic in *Figure 2.21*, a current which in prospect could reach 100 times the full load value will be interrupted in less than 0.001 s.

For overcurrent protection a device which will carry some overcurrent for a period without deterioration and which can be reset when it has operated is desirable. The relays associated with contactors are best suited to this function. Thermal and magnetic relays are cheaper than induction types and are used extensively for small motor protection. A series fuse with a rating slightly greater than the full load current protects the cable and contactor against short circuit faults. The fuse acts as back up protection to the relay and contactor. If an overload occurs and the relay fails to operate the fuse will clear the circuit.

The induction relay tends to be used for feeder protection.

DISTRIBUTION SYSTEMS

Radial systems

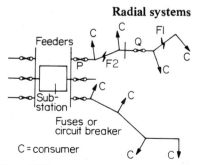

Feeders

Sub-station

Fuses or circuit breaker

C = consumer

Figure 2.22

A radial system for the distribution of electrical energy is shown in *Figure 2.22*. A substation supplies consumers C through radial distributors which fan out from the substation. A fuse or circuit breaker protects each distributor. Some distributors have subsidiary fuses which are of lower rating than the main fuse somewhere along their length. In the event of a fault on the feeder the relevant fuse clears leaving all consumers on that section without a supply. In *Figure 2.22* a fault F1 would cause fuse Q to clear leaving two consumers without a supply. A fault at F2 would cause fuse P to clear leaving all four consumers without a supply. Since there is no alternative method of supply to these consumers repairs to the line have to be carried out before they can be reconnected.

The consumer on the end of each distributor suffers voltage reductions as the load on that distributor increases and to minimise this the cross-sectional area of the conductors is large. This is therefore an expensive system to install and offers poor security of supply.

Fault finding is relatively simple since tests need only be done on the cleared section of line.

Ring system

Addition of further substations feeding the other ends of the radial feeders as shown in *Figure 2.23* effectively converts the system into a ring. This substantially reduces the voltage drops along the distributors and enables savings in conductor cross-sectional areas and costs to be made. A fault F1 would result in fuses Q and R clearing leaving two consumers without a supply. A fault F2 would result in fuses P and Q clearing and again only two consumers would lose their supply instead of four with the radial system.

At 11 kV a ring system employing isolators at each load point enables greater security of supply to be achieved. In *Figure 2.24* a fault F1 can be cleared by opening isolators I_2 and I_3 and no other section need be disconnected provided that a second fault does not occur. Meanwhile repairs can be carried out.

Interconnection of two points on the ring makes the system more versatile while reducing voltage drops and cable losses. This makes for greater expense however.

If circuit breakers are used instead of isolators, overcurrent relays which are sentisive to the direction of current flow can be used to automatically isolate a faulty section. When isolators are used there will be an interruption of supply to consumers while the fault is located and the isolators are manually operated.

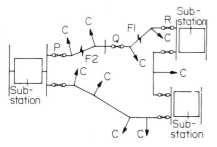

Figure 2.23

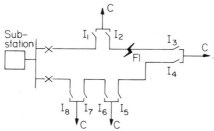

I = isolator (all isolators normally closed)   **Figure 2.24**

Reduction of voltage drops and cable losses together with increased security of supply are achieved by increasing the number of substations and the degree of interconnection. However, the expense rapidly increases with complexity and the actual arrangement employed is the best which can be obtained at a realistic cost to the consumer.

DISTRIBUTOR CALCULATIONS

Radial

In *Figure 2.6* we saw a single resistive load being fed from a 240 V supply. Using a single line representation as discussed in Chapter 1, the diagram can be re-drawn as shown in *Figure 2.25*. The resistance marked is that of both go and return conductors.

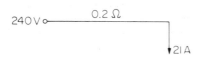

240 V $\circ$———0.2 Ω———┐

↓21 A

Figure 2.25

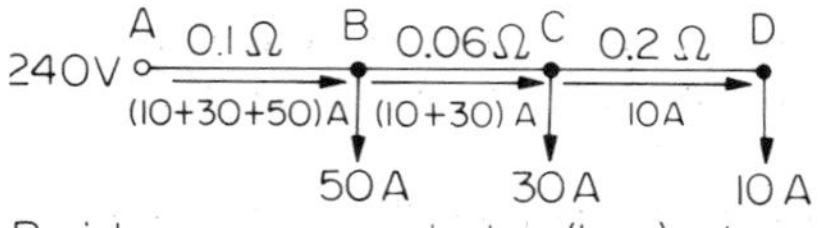

A 0.1 Ω   B 0.06 Ω   C 0.2 Ω   D
240 V $\circ$—•————•————•——↓
(10+30+50) A | (10+30) A | 10 A
      ↓   ↓   ↓
     50 A   30 A   10 A

Resistance are go and return (loop) values

Figure 2.26

The load voltage $= 240 - 21 \times 0.2$
$$= 240 - 4.2$$
$$= 235.8 \text{ V as previously.}$$

The method is suitable for both d.c. and single phase a.c. where resistance only is considered. When the loads or lines have inductance or capacitance there are phase angles to consider and arithmetic addition and subtraction of voltages gives incorrect results.

Let us now consider a radial feeder with three loads.

Example (1).   The details of a radial feeder are shown in *Figure 2.26.*

Calculate: (a) the load voltages
              (b) the power lost in the cable
              (c) the power developed by each load
              (d) the efficiency of the system.

Applying Kirchhoff's first law to each load point:
The current in section CD of the feeder = 10 A
Section BC carries load currents C and D = 30 + 10 = 40 A
Section AB carries the total load currents = 40 + 50 = 90 A
The voltage drop between A and B = $I_{AB} \times R_{AB} = 90 \times 0.1 = 9$ V
Voltage at load B = $240 - 9 = 231$ V
Power loss in section AB = $(I_{AB})^2 R_{AB} = 90^2 \times 0.1 = 810$ W
Power developed by load B = $V_B I_B = 231 \times 50 = 11\,550$ W
Repeat for section BC
Voltage drop from B to C = $40 \times 0.06 = 2.4$ V
Voltage at load C = $231 - 2.4 = 228.6$ V
Power loss in section BC = $40^2 \times 0.06 = 96$ W
Power developed by load C = $228.6 \times 30 = 6\,858$ W
Repeat for section CD
Voltage drop from C to D = $10 \times 0.2 = 2$ V
Voltage at load D = $228.6 - 2 = 226.6$ V
Power loss in section CD = $10^2 \times 0.2 = 20$ W
Power in load D = $226.6 \times 10 = 2266$ W
Total load powers = $2\,266 + 6\,858 + 11\,550 = 20\,674$ W
Total losses = $20 + 96 + 810 = 926$ W

$$\text{Efficiency} = \frac{\text{Power in loads}}{\text{Total power input}} = \frac{20\,674}{20\,674 + 926} \text{ OR } \frac{20\,674}{240 \times 90}$$
$$= 0.957 \text{ p.u.}$$

Distributor fed at both ends   Now consider the effect of reinforcing the system by feeding point D either from another substation or from the same substation through an extra length of cable or line.

Example (2).   Re-calculate (a) to (d) in Example 1 with the supply reinforced as shown in *Figure 2.27.*

In this case we firstly have to determine how much current is supplied from each end of the feeder.
Consider a current I_1 to be entering from the left hand end.
The current in the section AB = I_1 A and the volt drop between A and B = $I_1 \times 0.1$ V.

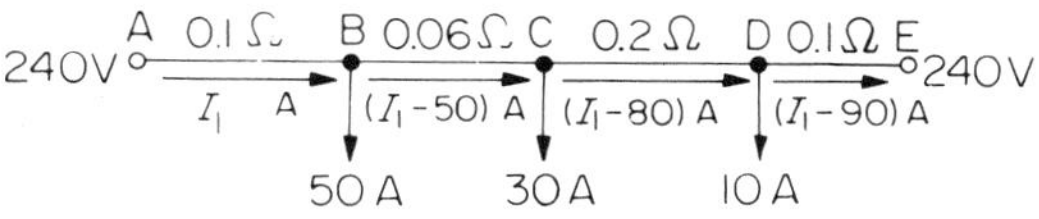

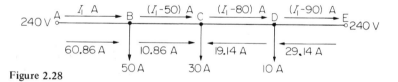

Figure 2.27

At point B a load current of 50 A is supplied so that the current flowing on in section BC must be $(I_1 - 50)$ A and the volt drop from B to C = $(I_1 - 50) \times 0.06$ V.

Similarly, the current in section CD = $(I_1 - 50) - 30 = (I_1 - 80)$ A and the volt drop from C to D = $(I_1 - 80) \times 0.2$ V.

In section DE the current is $(I_1 - 90)$ A and the volt drop from D to E = $(I_1 - 90) \times 0.1$ V.

The voltage at A — all the volt drops along the line = voltage at E. This is as in Example 1 except in this case we know the voltage at E = 240 V.

Therefore

$$240 - I_1 \times 0.1 - (I_1 - 50) \times 0.06 - (I_1 - 80) \times 0.2$$
$$- (I_1 - 90) \times 0.1 = 240$$

Multiply out the brackets

$$240 - 0.1I_1 - 0.06I_1 + 3 - 0.2I_1 + 16 - 0.1I_1 + 9 = 240$$

Transpose

$$240 - 240 + 3 + 16 + 9 = 0.1I_1 + 0.06I_1 + 0.2I_1 + 0.1I_1$$
$$28 = 0.46I_1$$
$$I_1 = 60.86 \text{ A}$$

Figure 2.28

Redraw the diagram as in *Figure 2.28.*

In section BC we have $(60.86 - 50)$ A = 10.86 A

In section CD we have $(60.86 - 80)$ A = −19.14 A. The negative sign indicates a reversal of current direction from that shown in *Figure 2.27* and this is seen to be logical since the 30 A load receives 10.86 A from end A and 19.14 A from end E.

Voltage at B = $240 - 60.86 \times 0.1 = 233.9$ V.

Power loss in section AB = $60.86^2 \times 0.1 = 370.4$ W.

Power in the load B = $233.9 \times 50 = 11\,695$ W.

Voltage at C = $233.9 - 10.86 \times 0.06 = 233.25$ V.

Power loss in section BC = $10.86^2 \times 0.06 = 7.08$ W.

Power in load C = $233.25 \times 30 = 6997.5$ W.

Voltage at point D. Since current flows from D to C, the voltage at D must be greater than that at C. The minimum voltage on the distributor is at load C and currents flow from both ends to this point.

Voltage at D = $233.25 + 19.14 \times 0.2 = 237.08$ V.

Power loss in section CD = $19.14^2 \times 0.2 = 73.26$ W.

Power in load D = 237.08 × 10 = 2370.8 W.
We know that the voltage at point E is 240 V. Let us check the calculations by adding the volt drop from E to D to the voltage at D

$$237.08 + 29.14 \times 0.1 = 240 \text{ V}$$

Power loss in section DE = $29.14^2 \times 0.1 = 84.91$ W.
Total load powers = 11 695 + 6997.5 + 2370.8 = 21 063.3 W.
Total losses = 370.4 + 7.08 + 73.26 + 84.91 = 535.65 W.

$$\text{Efficiency} = \frac{21\,063.3}{21\,063.3 + 535.65} = 0.975 \text{ p.u.}$$

Notice that by feeding the system at both ends each of the load voltages has been increased. The cable losses have been reduced so increasing the efficiency.

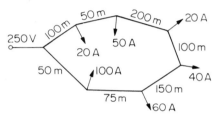

250 V
100 m
50 m
200 m
20 A
20 A
50 A
100 m
50 m
100 A
40 A
150 m
75 m
60 A
The resistance of 100m of single conductor
= 0.05 Ω

Figure 2.29

Example (3). For the ring main shown in *Figure 2.29*, determine the current in each section and the minimum load voltage.

Redraw the diagram putting in the go and return resistances. A 100 m go and return has a resistance of 2 × 0.05 = 0.1 Ω. A ring may be considered as a feeder fed at both ends at the same voltage.

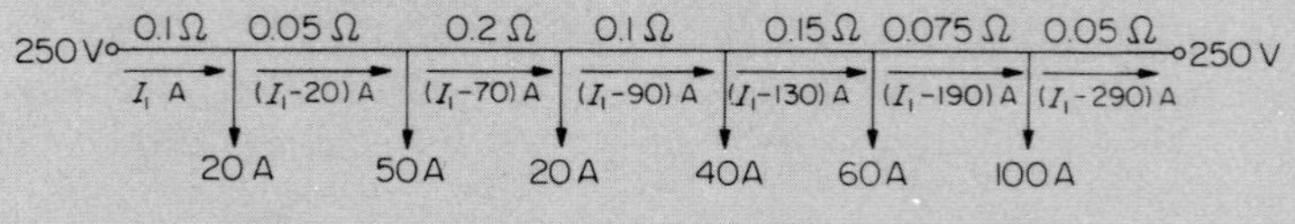

250 V
0.1 Ω
0.05 Ω
0.2 Ω
0.1 Ω
0.15 Ω
0.075 Ω
0.05 Ω
250 V
I₁ A
(I₁-20) A
(I₁-70) A
(I₁-90) A
(I₁-130) A
(I₁-190) A
(I₁-290) A
20 A
50 A
20 A
40 A
60 A
100 A

Figure 2.30

Starting with a current I_1 flowing towards the 20 A load.

$$250 - 0.1I_1 - 0.05(I_1 - 20) - 0.2(I_1 - 70) - 0.1(I_1 - 90)$$
$$- 0.15(I_1 - 130) - 0.075(I_1 - 190) - 0.05(I_1 - 290) = 250$$
$$250 - 0.1I_1 + 1 - 0.2I_1 + 14 - 0.1I_1 + 9 - 0.15I_1 + 19.5$$
$$- 0.075I_1 + 14.25 - 0.05I_1 + 14.5 = 250$$
$$72.25 = 0.725I_1$$
$$I_1 = 99.65 \text{ A}$$

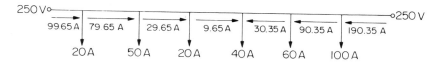

Figure 2.31

The minimum voltage occurs at the 40 A load. The voltage at this point may be determined starting at either end.
From the left
$250 - 99.65 \times 0.1 - 79.65 \times 0.05 - 29.65 \times 0.2 - 9.65 \times 0.1$
$= 229.16$ V
From the right
$250 - 190.35 \times 0.05 - 90.35 \times 0.075 - 30.35 \times 0.15 = 229.16$ V

PROBLEMS FOR SECTION 2

(4) The line voltage of a synchronous three-phase generator is 33 kV. What is the value of the voltage from one line to the neutral point?

(5) The output voltage from a three-phase transformer is 125 V per phase. What is the value of the line voltage?

(6) Why is care taken to balance loads on the phases of a three-phase system?

(7) What is the name given to an electrically operated switch used to control small motors?

(8) At what stage in the erection of a conduit wiring system are the conductors drawn in? Why is this?

(9) Give three limitations which must be considered when plastic conduit is used.

(10) What advantages has MICS cable over most other types of cable?

(11) What advantages has the HBC fuse over rewireable types?

(12) What is the reason for earthing (a) the star point of a three-phase supply transformer (b) the metal casing of equipment. Under what circumstances is the earth wire to consumers' equipment omitted?

(13) What advantages are claimed for Protective Multiple Earthing (P.M.E.) systems?

(14) What could be the outcome of high earth loop impedance on an installation?

(15) Why are gas and water pipes bonded to the supply neutral in a P.M.E. system?

(16) What advantage has a current balance earth leakage system over the H.B.C. fuse?

(17) A fuse with a fusing factor of 1.6 has a rating of 15 A. What is the minimum operating current? What does the rated value indicate?

(18) Why would one sometimes find an instantaneous relay fitted in series with a thermal relay in a motor circuit?

(19) Why would an H.B.C. fuse be used to back up relay protection?

(20) A radial feeder ABCD is fed at A at 200 V. The loads

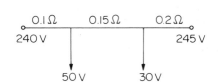

Figure 2.32

are 20 A, 10 A and 10 A at B, C and D respectively. The resistances of go and return conductors are:

AB = 0.1 Ω, BC = 0.15 Ω, CD = 0.05 Ω.

Calculate the efficiency of the system under these conditions.

(21) A feeder ABCD is fed at A and D at 220 V. A load of 20 A is situated at B which is 100 m from A. A load of 30 A is situated at C which is 120 m from D. The feeder is 420 m long. The resistance of 100 m of *single* conductor is 0.025 Ω. Determine the currents in each section of the feeder and the minimum voltage.

(22) For the feeder shown in *Figure 2.32* determine the power in the load with minimum potential.

(23) A ring main ABCDEFGA is fed at A at 250 V.

AB = 50 m, BC = 50 m, CD = 100 m, DE = 75 m, EF = 75 m, FG = 150 m and GA = 100 m.

100 m of single conductor has a resistance of 0.05 Ω. The loads are as follows:

B = 20 A, C = 30 A, D = 10 A, E = 50 A, F = 20 A and G = 25 A.

Determine the values of the currents in each section of the ring and the value of the minimum potential difference at a load.

3 Regulations

Aims: At the end of this section you should be able to:
Explain the need for, and discuss the scope of the *Regulations for the Electrical Equipment of Buildings, Electricity (Factories Act) Special Regulations* and the *Electricity Supply Regulations.*

SAFETY AND REGULATIONS

Inadequate control of electricity can give rise to serious dangers due to fire and shock. In the early days of the electricity public supply system the powers involved were small and faults were frequent often resulting in whole plants being shut down. At least one power stations was destroyed by fire when a fault developed which was not cleared. With the vast amounts of power involved today such a situation cannot be allowed to develop and over the years regulations applying to suppliers and consumers have been drawn up to prevent dangerous situations and accidents as far as is reasonably practicable. A major accident causes the regulations to be reviewed to see how best to prevent a future similar occurrence but it must be said that many of the original regulations were so well and widely cast that remarkably few changes have been required over the years.

These regulations are:

The Electricity Supply Regulations.
The Electricity (Factories Act) Special Regulations.
Regulations for the Electrical Equipment of Buildings, drawn up by *The Institution of Electrical Engineers.*

In addition there are special regulations applying to coal mines, cinemas, places of public entertainment and petroleum installations.

THE ELECTRICITY SUPPLY REGULATIONS

These are issued by the Department of Trade and Industry. They are designed to secure the safety of the public and to ensure a proper and sufficient supply of electrical energy. They apply to the electricity distribution system covering construction, operation, protection, maintenance and safety up to the consumers' terminals.

Certain sections relate to the consumers' installation in as much as they give the Area Boards powers to insist on certain standards of work in these installations before the supply can be connected. The supply may be withdrawn from installations which subsequently become unsafe. The standards generally required are laid down in the Regulations for the Electrical Equipment of Buildings. In addition a supply may be refused to a consumer with equipment likely to adversely affect other consumers.

The Area Board has to declare the voltage of the supply to the consumer and to maintain this within 6 per cent of that value.

The frequency has to be maintained within specified limits. This is under the control of the Generating Board and except in very rare circumstances they maintain an average of 50 cycles per second (Hertz) over the day so keeping electric clocks correct.

Both Area and Generating Boards have very strict sets of safety rules and operate a 'Permit-to-work' system which ensures an extremely low accident rate.

ELECTRICITY (FACTORIES ACT) SPECIAL REGULATIONS

These deal with factory installations and the operation of electrical plant. They have been drawn up to ensure safe working in factories and are necessary to prevent unsafe installations and practices which might otherwise be employed either out of ignorance or for cost saving. The Health and Safety at Work Act 1974 places responsibility for safe working both on management and employee.

The Factory Inspectorate can issue improvement notices in respect of substandard equipment or method of working requiring them to be brought up to standard in a certain period of time. In extreme cases prohibition notices are issued which take effect immediately. Prosecution can result from contravention of these notices and in accident cases where it can be proved that the accident has been caused by a contravention of the regulations.

There are 32 regulations in all and some of these are now quoted in part showing how they promote safe working in particular circumstances. It follows that non-compliance will create a hazardous or unsafe condition.

Fixed apparatus

Reg. 11. Every motor, convertor and transformer shall be protected by efficient means suitably placed, and connected so that all pressure may thereby be cut off from. . .the apparatus. . .provided that where one point of the system is earthed there shall be no obligation on that side of the system which is connected to earth.

Reg. 12. Every electrical motor shall be controlled by an efficient switch or switches for starting and stopping so placed as to be easily worked by the person in charge of the motor. In every place in which machines are being driven by any electric motor there shall be means at hand for either switching off or stopping the machines if necessary to prevent danger.

Reg. 21. Where necessary to prevent danger adequate precautions shall be taken either by earthing or by other suitable means to prevent any metal other than the conductor from becoming electrically charged.

Reg. 25. Adequate working space and means of access free from danger shall be provided for all apparatus that has to be worked or attended to by any person.

Reg. 26. All those parts of premises in which apparatus is placed shall be adequately lighted to prevent danger.

In addition to these regulations the general requirements of the Factories Acts governing moving equipment apply. These state that apparatus which has dangerous moving parts must be securely fenced or be safe by virtue of its position.

SUBSTATIONS AND TRANSFORMERS

A substation is defined as '. . .any premises or that part of any premises in which energy is transformed or converted from or to a pressure above medium pressure except for purposes of working instruments, relays or similar auxiliary apparatus, if such premises or part of premises are large enough for a person to enter after the apparatus is in position.'

Reg. 30. Every substation shall be substantially constructed and shall be so arranged that no person other than an Authorised person can obtain access thereto otherwise than by the proper entrance or can interfere with the apparatus or conductors therein from outside; and shall be provided with an efficient means of ventilation and be kept dry.

Reg. 31. Every substation shall be under the control of an Authorised person and none but the Authorised person or person acting under his immediate supervision shall enter any part thereof where there may be danger.

Reg. 32. Every underground substation shall be provided with adequate means of access by door or trapdoor with a staircase or ladder securely fixed so that no live part shall be within reach of a person thereon.

Reg. 20. Where a high pressure or extra-high pressure supply is transformed for use at a lower pressure or energy is transformed up to above low pressure suitable provision shall be made to guard against danger by reason of the lower-pressure system becoming accidentally charged above its normal pressure by leakage or contact from the higher voltage system.

Permanent connection of one part of the lower voltage system to earth is the best method of complying with this regulation and in the case of the public supply the provision is generally made by the Supply Authority under the Electricity Supply Regulations. (The neutral point is earthed).

In addition to the regulations there are three pages of definitions which are generally self-explanatory but for a full understanding of Regulation 20 these relevant pressures are quoted here.

'Pressure' means the difference of electrical potential between two conductors or between conductor and earth as measured using an r.m.s. voltmeter.

'Low pressure' is up to 250 V.

'Medium pressure' is between 250 V and 650 V.

'High pressure' is between 650 V and 3000 V.

'Extra-high pressure' is upwards of 4000 V.

THE INSTITUTION OF ELECTRICAL ENGINEERS' REGULATIONS FOR THE ELECTRICAL EQUIPMENT OF BUILDINGS

The I.E.E. Regulations are designed '. . .to ensure safety, especially from fire and shock, in the utilisation of electricity in and about buildings'.

They relate to consumers' installations and specify conductor and cable types, methods of installation or wiring and apparatus, safety measures and testing methods.

'Consumer's installation' is defined as wiring and apparatus situated in the consumer's premises and controlled by him excluding any switch-gear of the supply undertaking which the consumer is permitted to use.

A regulation common to the Electricity Special Regulations (Reg. 1) and the I.E.E. Regulations (Reg. 1) states that 'All apparatus and conductors shall be sufficient in size and power for the work they are called upon to do and shall be so constructed, installed and protected so as to prevent danger so far as is reasonably practicable'.

This states basically the object of all the regulations.

The I.E.E. Regulations are not legally enforceable as in the case of the Factories Act Special Regulations but almost certainly work which did not comply with the I.E.E. Regulations would contravene the Factories Act. The Factories Act often calls for additional requirements in fact. Also under the Supply Regulations the Area Boards are empowered to refuse a supply to substandard installations and the standard is generally that set by the I.E.E. Regulations.

Certain installations where there are exceptional risks must also comply with additional regulations namely those for coal mines, cinemas etc., as previously mentioned.

Tables of cable sizes for different currents are provided together with factors for close and coarse protection as mentioned in Chapter 2.

The Regulations are updated from time to time as new materials and new systems become available.

COMPARISON BETWEEN THE SCOPES OF THE REGULATIONS

The Supply Regulations apply up to the supply point in the factory or premises. The I.E.E. Regulations relate to installation practice on consumers' premises. Work done to the I.E.E. Regulations is of high enough standard to satisfy the Supply Regulations and to a considerable extent the Factories Act.

The Factories Act has the force of law behind it. Prosecution can result from non-compliance and the I.E.E. Regulations could be quoted to demonstrate malpractice.

The I.E.E. Regulations are likely to be the basis of a contract between an electrical installation firm and the purchaser of an installation.

PROBLEMS FOR SECTION 3

(1) What is the general objective of all electricity supply and wiring regulations?

(2) What powers do the Electricity Supply Regulations give to the Area Boards with respect to consumers' equipment?

(3) What do the Electricity Supply Regulations have to say about terminal voltage?

(4) What additional safety measures are taken in the Area and Generating Boards' own premises to minimise accident rate?

(5) What two measures are open to Factory Inspectors within factories if unsafe working practices or machines are found?

(6) What is the difference in scope between the Factories Act and the I.E.E. Regulations?

(7) What is the basic objective of the I.E.E. Regulations?

(8) What additional regulations may have to be complied with in certain hazardous situations?

(9) In a court case involving the Factories Act, what weight might the I.E.E. Regulations have?

(10) Which of the regulations/acts mentioned apply to consumers of electrical energy?

(11) Which of the regulations/acts mentioned apply to the suppliers of electrical energy?

4 Tariffs and power factor correction

Aims: At the end of this section you should be able to:
Define maximum demand, diversity factor, load factor and power factor.
Explain the need for and the use of a two part tariff.
Calculate the charges for electrical energy on various tariffs.
Calculate the savings to an industrial consumer brought about by an
improvement in power factor.

GENERATION AND TRANS-MISSION COSTS

The cost of generation and transmission of electrical energy is divided into two parts:
1. Capital charges
2. Running charges.

Capital charges

In order to build a power station or a transmission line money is borrowed and interest paid annually. Money is also put aside so that in theory at least, at the end of 25 or 30 years the loan can be repaid. This is called a depreciation allowance and with an expanding system is in fact spent more or less continuously on new plant. Both interest and depreciation charges have to be met whether the plant purchased is used or not.

Running charges

In order to run a power station men must be paid, fuel purchased and repairs carried out. On overhead lines and underground cables repairs must be done and routine testing and inspections carried out. These costs are very nearly proportional to the amount of energy sold. The cost of electrical losses in the system are also included in the running charges.

In the UK electricity is purchased from the Generating Board by the Area Electricity Boards who act as retailers to industry and the general public. The Area Boards and all their consumers must in turn pay both the capital and running charges of the system and therefore all tariffs have a two part structure; a capital charge which is a sum based on the availability of plant to meet the consumers' demand, generators, lines and transformers; and a running charge which is based on the cost of fuel and other resources used in the production of the energy consumed. The running charge is fixed at intervals but the Generating and Area Boards have powers to vary this automatically as fuel costs change by means of a 'Fuel Adjustment Charge'.

For example: cost per kWh = 2.2 p + 0.000 08 p for every penny average fuel costs in the region increase above £15 per tonne.

Very large consumers may be supplied at 132 kV or 33 kV. A supply at these voltages does not involve the use of the rest of the distribution system down to 415/240 V and since less plant is involved there will be a saving in capital charges. In addition there will be a reduction in system losses enabling a slight reduction in running charges to be made.

POWER FACTOR

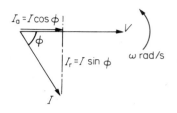

Figure 4.1

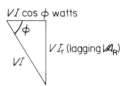

Figure 4.2

Most loads on the electricity supply system comprise resistance and inductance in series so that the supply current lags on the voltage by an angle ϕ° as shown in *Figure 4.1*.

The amount of this current which is in phase with the voltage and therefore capable of doing work is called the active component of current I_a.

The power in the circuit = VI_a watts
The product $V \times I$ is termed volt-amperes.

$\dfrac{\text{True power}}{VI}$ is the circuit power factor.

Hence power factor = $\dfrac{VI_a}{VI} = \dfrac{I_a}{I}$ which from *Figure 4.1* can be seen to be equal to $\cos \phi$.

Also in *Figure 4.1*, the vertical side of the current triangle I_r can be found since $I_r/I = \sin \phi$ so that $I_r = I \sin \phi$.

In a circuit comprising pure resistance, all the current is in phase with the voltage so that $I = I_a$ and $I_r = 0$. I_r is only present when the circuit has reactance and is therefore known as reactive current.

Taking the current triangle in *Figure 4.1*, multiplying each side by V, the circuit voltage, gives a similar triangle showing the relationship between power, volt-amperes and volt-amperes-reactive as the product $V \times I_r$ is called.

Since the current lags on the voltage in this case these are known as lagging volt-amperes-reactive.

[Notice carefully in *Figure 4.2*: I is the symbol for current whilst A is the unit of current. We write $I = 5$ A for example. Similarly $VI_r = 250\ VA_R$]

MAXIMUM DEMAND

The Board must obtain payment from its consumers for the amount of plant involved in its operation irrespective of the number of hours that it runs. Consider a small system of say 100 000 kW capacity, the total capital charges on which are £1.5 million per annum. This represents £15 for each kilowatt of plant capacity per annum. 100 000 consumers each with a 1 kW simultaneous demand should each pay £15 per annum to cover the capital charges. Whether the consumer leaves the equipment on for the whole year or just for the one hour the amount of plant involved is the same and the charge is the same. There would of course be a difference in the energy or running charge between the two cases.

When the Area Boards purchase energy from the Generating Board or industrial consumers from the Area Boards the amount of plant involved is determined by measuring maximum demand. This may be on a kW or kVA basis. In the former case a kilowatt-hour meter has its advance measured during each half hour of the year. A clock times half hours from the hour to half past and from half past to the hour. The number of kWh used in one half hour multiplied by two gives the hourly rate in kWh per hour. [kWh/h = kW]

The largest value of the kW demand in a given period (month, quarter or year) is the maximum demand for the period. Some recorders print the demand each half hour on a paper tape whilst others move a pointer round a scale leaving it at the highest point reached.

Where charges are based on kVA it is necessary to take power factor into account. There are a number of ways of doing this amongst which is to record both kWh and kVA_Rh from which the kVA and power factor can be computed.

DIVERSITY FACTOR   Within a factory or premises, not all the installed equipment will be working simultaneously.

$$\text{Diversity factor} = \frac{\text{Demand of equipment actually connected at any instant}}{\text{Maximum demand}}$$

$$\therefore MD = \frac{\text{TOTAL LOAD}}{D.F.}$$

LOAD FACTOR

$$\text{Load factor} = \frac{\text{Energy consumed in a given period}}{\text{Energy that would have been consumed had the maximum demand been sustained during that period}}$$

COST OF ELECTRICAL ENERGY

Domestic   Because of the high cost of special metering and the relatively low demand involved the capital charges are recovered from domestic consumers by using a front end loaded tariff with possibly a fixed charge in addition. The use of this type of tariff is best illustrated using an example.

Example (1).   Calculate the cost of electricity supplied to domestic premises which have a load factor of 0.05 (5%) and a maximum demand of 10 kW for the following two tariffs:
(a) Fixed charge £5; First 150 kWh at 6 p/kWh, all over 150 kWh 2.5 p/kWh. All charges per quarter which is 91 days.
(b) Flat rate of 4 p/kWh

(a) Load factor = $\dfrac{\text{Energy consumed per quarter}}{\text{Maximum demand} \times \text{hours per quarter}}$

$$0.05 = \frac{\text{Energy consumed}}{10 \times 91 \times 24}$$

Energy consumed = 1 092 kWh per quarter.
Fixed charge = £5
150 kWh at 6 p/kWh = £9
(1092 − 150) kWh at 2.5 p/kWh = £23.55

Total cost = £37.55   Average price per kWh = $\dfrac{3\,755}{1\,092}$

$$= 3.44 \text{ p/kWh.}$$

(b) 1 092 kWh at 4 p/kWh = £43.68.

Example (2).   A family uses 2000 kWh in a winter quarter. The quarterly tariff is:
First 100 kWh cost 8 p/kWh: all over 100 kWh cost 2.4 p/kWh. The maximum demand is 12 kW.
Calculate: (i) the average cost per kWh
(ii) the load factor for the quarter.

Industrial A number of variations on tariffs exist and these are best illustrated using a further example.

Example (3). A factory has a maximum demand of 200 kW, a load factor of 0.4 (40%) and an average operating power factor of 0.7.

Calculate the annual cost of electrical energy on the following three tariffs:
(a) Maximum demand charge £15/kW. Running charge 1.8 p/kWh.
(b) Maximum demand charge £12/kVA. Running charge 1.8 p/kWh.
(c) Basic maximum demand charge £15/kW increased by a factor 0.1 (10%) for each 0.1 that the power factor is worse than 0.9. Running charge 1.8 p/kWh.

$$(a)\ 0.4 = \frac{\text{Energy consumed}}{\text{Maximum demand} \times \text{hours in the year}}$$

Energy consumed = 0.4 × 200 × 365 × 24 = 700 800 kWh.
Maximum demand charge = £15 × 200 = £3 000.
Energy charge = 700 800 × 1.8 p = £12 614.
Total cost = £15 614.
[Average price per kWh = 2.228 p]
(b) Power factor = 0.7 = cos ϕ (see *Figure 4.3*)

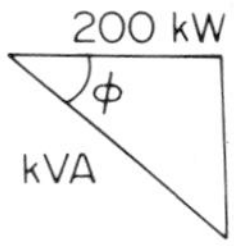

$$\frac{200}{\text{kVA}} = 0.7 \quad \text{Maximum demand in kVA} = \frac{200}{0.7} = 285.71$$

Maximum demand charge = £12 × 285.71 = £3 428.52.
Energy charge as before = £12 614.
Total cost = £16 042.52.
[Average price per kWh = 2.289 p]
(c) The power factor = 0.7. This is 0.2 worse than 0.9 so that the maximum demand charge is increased by a factor of 0.2.
The maximum demand cost = £15 + 0.2 × £15 = £18 per kW.
Maximum demand charge = £18 × 200 = £3 600.
Energy cost = £12 614.
Total cost = £16 214.
[Average cost per kWh = 2.314 p.]

Example (4). A factory with an annual consumption of 1 million kWh has a load factor of 0.5. The electricity tariff is £14 per kW of maximum demand plus 1.7 p/kWh. Calculate the average cost per kWh.

Figure 4.3

Effect on electricity charges of improving load factor. As an example suppose a factory has an energy requirement of 1 million kWh per annum.
The electricity tariff has a maximum demand charge of £15/kW.
Let us consider the effect of changing load factors on the maximum demand charge.
(a) With a load factor of 0.25

$$0.25 = \frac{1\ 000\ 000}{\text{Max. demand} \times 365 \times 24}$$

Maximum demand = 456.6 kW
Annual maximum demand charge = £15 × 456.6 = £6 849.31
(b) With a load factor 0.75
Using a similar calculation to that above, the maximum demand
    = 152.21 kW
Maximum demand charge = £2 283.1
    Saving = £4 566.21
The energy costs will be the same in both cases.

It can be seen therefore that an improvement in the load factor leads to cost saving where the tariff includes a maximum demand charge.

This improvement in load factor may be achieved in a number of ways. Suppose a factory with a metal foundry and a machine shop both started up simultaneously in the morning. There will be a heavy demand for the first hour or so after which the demand will reduce as furnaces reach working temperature, motors warm up and in winter the workshops reach a comfortable temperature and possibly lighting is switched off. The load factor is low since the heavy demand only lasts for a small proportion of the working day.

By either working a night shift in the foundry or starting it up say two hours earlier than the machine shop, the demand by the foundry can be reduced to a very low value when the machine shop starts running. The maximum demand is thereby reduced and the load factor increased. The total energy used remains at about the same level.

The load factor on commercial premises could be improved by using off peak storage heating instead of direct acting equipment so that the heating load is off when lighting and other services are required for the morning start.

Increasing the number of hours that a particular piece of equipment works improves its load factor. Where several pieces of equipment are used only intermittently ensuring that only one is used at a time improves the load factor.

Effect of power factor improvement.   Where a tariff incorporates a maximum demand charge based on kVA, or has a power factor penalty clause as in Example 3(c), savings can be achieved by improving a low power factor. Such a tariff seeks to discourage low power factor since this involves larger currents than necessary to perform a given amount of work. Conductors, transformers and switchgear must therefore be larger than for the same load at unity power factor.

Consider a factory with a maximum demand of 400 kW paying for electricity on a tariff of £12/kVA of maximum demand.

(i) At a power factor of 0.2 lagging. ϕ_1 = 78.46° (see *Figure 4.4*)

$$kVA_1 = \frac{400}{0.2} = 2\,000$$

Maximum demand charge = £12 × 2 000 = £24 000

(ii) At a power factor of 0.6 lagging. ϕ_2 = 53.1°

$$kVA_2 = \frac{400}{0.6} = 666.67$$

Maximum demand charge = £12 × 666.67 = £8 000

(iii) At a power factor of 0.9 lagging. ϕ_3 = 25.8°

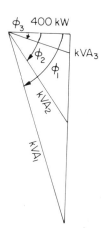

Figure 4.4

$$kVA_3 = \frac{400}{0.9} = 444.4$$

Maximum demand charge = £12 × 444.4 = £5 333.3

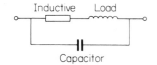

Figure 4.5

Thus considerable savings can be made by improving the operating power factor. The improvement may be brought about by the use of capacitors connected across the supply either at the load point itself or in the substation feeding the factory.

A capacitor draws a current which leads on the supply voltage by $90°$ and therefore has no active component. It is totally reactive and multiplying by the supply voltage V gives VI_c, leading volt-amperes reactive (see *Figure 4.6*).

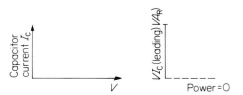

Figure 4.6

Adding the capacitor to the load does not affect the load power but since the leading and lagging volt-amperes reactive are in direct phase opposition the arithmetic difference may be taken as shown in *Figure 4.7*.

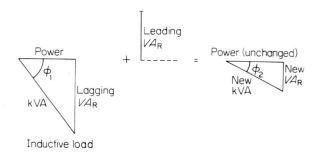

Figure 4.7

New value of VA_R = Lagging VA_R − Leading VA_R
The overall phase angle has been reduced from ϕ_1 to ϕ_2 and the value of kVA has been reduced.
As well as capacitors, there are specially designed motors which can be caused to operate at leading power factors so having the same effect. These are installed to drive loads within the factory.

Example (5). An industrial load has a maximum demand of 300 kW at a power factor of 0.6 lagging. Calculate the saving in maximum demand charges and the overall saving if a capacitor is

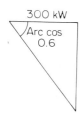

Figure 4.8

Figure 4.9

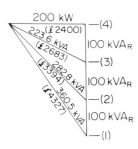

Figure 4.10

fitted which draws 150 kVA$_R$. The tariff is £12/kVA of maximum demand. The annual capital charges for the capacitor are £600.

[Note that capacitors, like all plant, have to pay capital charges as already discussed. For a capacitor the running charges are considered to be zero since there is virtually no power loss]

cos ϕ = 0.6 therefore ϕ = 53.13° (using tables or calculator)

$$kVA = \frac{300}{0.6} = 500$$

$$\frac{kVA_R}{300} = \tan \phi = 1.33. \quad kVA_R = 1.33 \times 300 = 400$$
$$\text{(or use Pythagoras' theorem)}$$

Maximum demand charge = £12 × 500 = £6 000.
Since the capacitor draws 150 kVA$_R$ leading, the total number of lagging kVA$_R$ will be reduced by this amount to 400 − 150 = 250 kVA$_R$. Redraw the triangle as in *Figure 4.9*.

By Pythagoras' theorem: kVA = $\sqrt{(300^2 + 250^2)}$ = 390.5.
New maximum demand charge = £12 × 390.5 = £4 686.
Saving on maximum demand charge = £6 000 − £4 686 = £1 314.
But the annual cost of the capacitor = £600. Therefore the net saving = £1 314 − £600 = £714.

Example (6).   Calculate the saving in maximum demand charge if a factory with maximum demand 500 kW at a power factor of 0.7 lagging improves this to 0.9 lagging. The maximum demand charge is £14/kVA.
What is the overall annual saving if the power factor correction equipment costs £800 per annum?

It is usually uneconomic to correct the power factor to unity since the costs of so doing become greater than the savings as the power factor approaches this value.

Consider the case of a factory with a maximum demand of 200 kW at a power factor such that it requires 300 kVA$_R$ lagging. This is condition (1) in *Figure 4.10*.

We will consider the effects of adding capacitor banks to the system in three stages each drawing 100 kVA$_R$ leading. The capital charges on each 100 kVA$_R$ bank are £400 per annum. The maximum demand charge is £12/kVA.

With no capacitors in circuit the maximum demand is 360.5 kVA and the maximum demand charge is £12 × 360.5 = £4 327.

Adding one bank of capacitors improves the power factor so reducing the kVA demand to 282.8 at a cost of £3 394. (point (2) in *Figure 4.10*).

Adding the next bank of capacitors reduces the kVA demand to 223.6 at a cost of £2 683. (point (3) in *Figure 4.10*).

Adding the final capacitor bank brings the power factor to unity when the kVA demand is equal to the number of kW and the maximum demand charge is £12 × 200 = £2 400. (point (4) in *Figure 4.10*).

The savings from (1) to point (2) = £933 at a cost of £400 spent in capacitors.

net saving = £533.

From point (1) to point (3) the savings = £1 644 at a cost of £800.

net saving = £844.

Correcting to unity, from point (1) to point (4) the saving in maximum demand charge is £1 927 at a cost of £1 200. Net saving £727.

Alternatively consider changing from point (3) to point (4). The saving in maximum demand charge = £283 which costs £400 for capacitors.

Figure 4.11 shows a graph of savings against power factors for the above case. The maximum savings are achieved by correcting to a power factor of 0.942. Going beyond this value costs more for capacitors than can be saved on maximum demand charges. Different capacitor and maximum demand charges will alter the power factor for optimum savings.

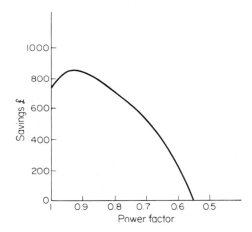

Figure 4.11

PROBLEMS FOR SECTION 4

(7) A private house contains the following equipment:
Electric lamps, total power 1100 W.
Television set 200 W
Electric iron 750 W
Two electric fires 4000 W
Washing machine with heater 4500 W
Food mixer 500 W
Electric cooker 10 000 W
Electric kettle 3000 W

(a) Calculate the diversity factor when (i) 200 W of lighting, the television set and both electric fires are operating, (ii) the cooker is operating at half maximum power, the electric iron and kettle are switched on.

(b) Calculate the load factor for the installation if in one quarter year 1250 kWh were consumed and the maximum demand was that of 50% of the total equipment listed above.

(c) Calculate the total electricity charge for the quarter and the average cost per kWh for a tariff: Fixed charge £2.50, First 150 kWh at 2.8 p/kWh, all in excess of 150 kWh, 2.1 p/kWh.

(8) The basic charge per kWh on a certain tariff is 2.1 p. There is a fuel adjustment clause whereby the unit charge may be increased by 0.0001 p for each penny that fuel cost rises above £18 per tonne. Calculate the cost per kWh when fuel costs £22 per tonne.

(9) During eight successive half hours the energy meter connected to the supply lines to a commercial premises was noted to make the following advances:
100, 150, 125, 175, 200, 150, 125 and 130 kWh respectively.
Assuming that these are the highest noted during the year, determine the maximum demand charge on a tariff of £14/kW of maximum demand.

(10) Why is there no cost advantage for domestic consumers to practise (i) power factor improvement or (ii) load factor improvement?

(11) A factory consumes 1.2 million kWh per annum. It has a load factor of 45% and operates at an average power factor of 0.65 lagging.

(a) Calculate the total electricity charges on a tariff of £14/kVA of maximum demand plus 1.6 p/kWh.

(b) Calculate the savings to be made by improving the power factor to 0.9 lagging. The capital charges on power factor correction equipment are £5/kVA$_R$.

(c) Calculate the additional saving in maximum demand charge by improving the power factor to unity. What additional cost is incurred in the provision of correcting equipment?

Would this final improvement be economically sound?

5 Materials and their applications in the electrical industry

Aims: At the end of this section you should be able to:
Compare aluminium and copper as conductor materials.
State insulating materials for circuit boards, cables and overhead lines.
Explain how flashover can occur.
Explain why the current flowing in the event of short circuit fault must be limited.
Explain the effect of ambient temperature on the current-carrying capacity of cables.
Use cable derating factors.
Describe the construction of power and light current cables.
Describe the construction of a printed circuit board.

CONDUCTOR MATERIALS FOR OVERHEAD LINES AND UNDERGROUND CABLES

The best electrical conductor known is silver but this is far too expensive and rare to provide all the conductor material required by the electrical industry. Next in order of conductivity come copper and and aluminium and these are the most important current carrying materials used in cable and line manufacture. Other materials used are cadmium-copper, phosphor-bronze and for some high-voltage low-power links, galvanised steel.

The conductivity of both copper and aluminium falls very rapidly with very small additions of alloying elements so that they are generally used pure. In the case of aluminium the mechanical strength is improved by using a stranded conductor with steel strands at the centre.

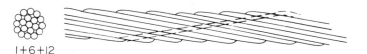

1+6+12

Figure 5.1

Figure 5.1 shows a stranded conductor and the make up will be 1 strand of steel plus 6 strands of aluminium; 7 strands of steel plus 12 strands of aluminium, or by adding a further layer of aluminium strands, 7 of steel plus 30 of aluminium.

The conductivity of the whole is taken as that of the aluminium alone since steel has a very high impedance and the current flows almost exclusively in the aluminium.

The total strength of the cable is normally 50% greater than that of the same conductivity copper cable. The result of using steel cores is to produce cables which are smaller for a given tensile strength than copper or, although larger for a given resistance, much stronger than copper. The table details some of the electrical and mechanical properties of copper and aluminium.

Property	Copper	Aluminium
Weight	87 200 N/m³	26 700 N/m³ (Aluminium 0.306 times copper)
Resistivity	1.73×10^{-8} Ωm (0.975 × silver)	2.87×10^{-8} Ωm (Aluminium 1.64 times copper) (0.585 × silver)
Strength	Ultimate 320 MN/m²	Ultimate (Aluminium 0.45 times copper) 144 MN/m²
Flexibility	When annealed, quite good. Used hard for cables and overhead lines when it must be stranded to give the required flexibility.	Very flexible, can be used solid in cables.
Jointing	Soldered ferrules for cables. Compression joint on overhead lines.	Compression type on overhead lines. Crimped lugs for cable terminations. Can be soldered or welded using special fluxes but generally more difficult than copper.
Resistance to corrosion	Excellent. Virtually none in most circumstances.	Poor when in contact with other metals and in particular copper and copper bearing alloys. Special sleeves and fittings required. On overhead lines there is limited deterioration, much of the original 1933 grid is still operational in its original form.

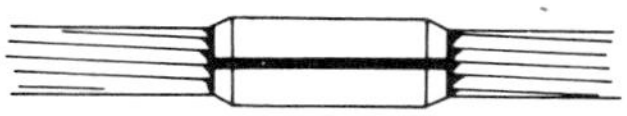

Cables joined using split ferrule and soft solder

Compression sleeve joining overhead lines

Crimped terminal on cable end

Figure 5.2

INSULATING MATERIALS FOR CABLES

For cables used up to 3.3 kV, the most common insulating material is polyvinyl chloride, (PVC). Street mains have either copper or aluminium conductors and are PVC insulated, steel wire armoured and covered overall with PVC to prevent corrosion.

House wiring usually comprises copper conductors which are insulated with PVC.

Polychloroprene (PCP) is used to insulate copper conductors in farming installations where resistance to ammonia and other such corrosives is required.

The insulating material in mineral insulated cables is magnesium carbonate which spaces solid copper conductors within a copper tube. It has a number of advantages amongst which is the ability to operate at red heat; a property which no other cable insulating material can match.

Vulcanised Indiarubber is now rarely used but may be found in older installations.

At 11 kV some polythene insulation is found but in the main paper is used at this voltage level. At voltages above 11 kV paper insulation with oil or gas filling predominates.

CABLE CONSTRUCTION

At low and medium voltages the single stranded conductor with PVC insulation is most commonly used. These are drawn into steel or plastic conduit (*Figure 5.3a*).

In domestic and commercial situations the twin copper conductor with an earth wire, PVC insulated, is often used (*Figure 5.3b*). These cables are either run on the surface or buried in the walls. In hazardous situations, mostly in industry, the mineral insulated cable is found

(MICS) and this is also run on the surface or buried (*Figure 2.11*).

For connecting appliances to the supply at low voltages via a plug and socket, the three-core flexible cable is employed. This comprises fine stranded copper wire which withstands flexing, PVC insulation and belt, covered overall with a braided protective fibre or fabric sheath.

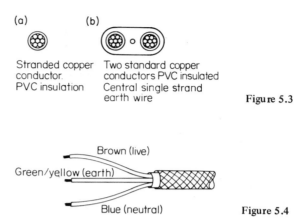

(a)

Stranded copper
conductor.
PVC insulation

(b)

Two standard copper
conductors PVC insulated
Central single strand
earth wire

Figure 5.3

Brown (live)

Green/yellow (earth)

Blue (neutral)

Figure 5.4

HEATING OF CABLES

Since all conductors in normal service have resistance, the passage of current along the conductor gives rise to a voltage drop. The voltage drop is in phase with the current producing it so that the power loss is equal to the product of voltage drop and current flowing.

Consider one core: Volt drop = IR where R = resistance of one core. Power loss = volt drop × current = $IR \times I = I^2 R$ watts per core.

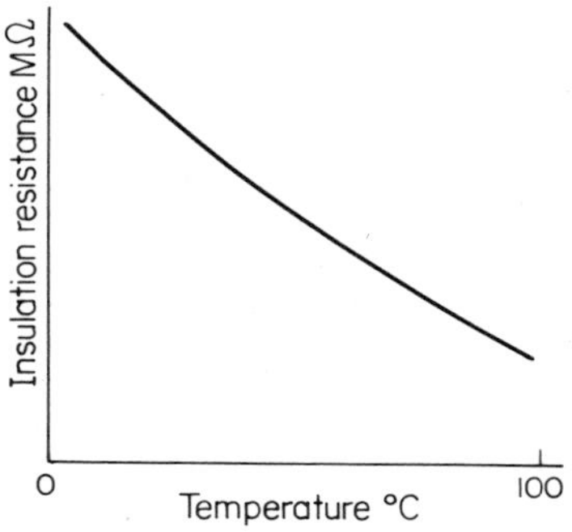

This power heats the core producing a rise in temperature of the conductor and hence of its insulation. The heat is lost to the surrounding air or ground. The temperature rises until the rate of dissipation from the insulation is equal to the rate of heat production in the conductor. Unfortunately good electrical insulators are often good heat insulators and it is quite possible to produce heat in the conductor at such a rate that the temperature at which equilibrium would be reached would also damage the insulation. In addition, the leakage current through the insulation must be considered. This produces heat as does the current flowing in the conductor. For example a cable operating at 240 V which has an insulation resistance of 240 megohms will allow $240/240 \times 10^6$ amperes to flow and the power loss will be 0.24 mW.

Figure 5.5 shows that as the temperature of a cable insulating material rises its insulation resistance falls. Hence at a higher temperature the leakage current is greater so that more heat is produced in the insulation. An unstable situation can be created where more heat increases the temperature which lowers the resistance of the insulation so allowing more leakage current to flow which produces more heat etc. The cable eventually fails by burning under these circumstances.

PVC becomes soft at temperatures in excess of 80°C and the conductors tend to migrate through the insulation eventually touching each other or earth. An upper working temperature limit of 70°C for PVC is set for this reason.

Insulation resistance MΩ

Temperature °C

0   100

Figure 5.5

EFFECT OF AMBIENT TEMPERATURE ON RATING

The amount of heat which can be conducted away from a cable depends on the temperature difference between it and the surrounding medium. A conductor which is at 60°C in an environment which is at 60°C cannot dissipate any heat. The same conductor in an environment at 5°C will dissipate a large amount of heat.

The upper working temperature of the insulation is fixed at a value at which it will not deteriorate rapidly and at which its mechanical properties are unimpaired. In the previous section we saw that this was 70° for PVC. The heat which can be dissipated and hence the maximum permissible current depends therefore on the temperature of the surrounding medium or ambient temperature as it is called.

A cable which can safely carry 20 amperes when the ambient temperature is 25°C may only be able to carry 12 amperes when the ambient temperature is 60°C. Tables of derating factors are contained in the I.E.E. regulations (see Chapter 3).

CABLES IN DUCTS OR IN CLOSE PROXIMITY

Cables which are in ducts or are laid close to each other suffer mutual heating. One cable becoming warm heats its neighbours. Where cables touch each other the number of free paths by which heat can escape to the environment is reduced. The general mass of cables may raise the temperature of the surrounding air or ground. For these reasons it is necessary to reduce the current rating of the conductors. A derating factor is applied which may be determined by experiment or obtained from sources such as the I.E.E. regulations or Electrical Research Association data.

For example from Table 1M of the I.E.E. wiring tables: A copper conductor 1 mm^2 cross-sectional area with PVC insulation has a current rating of 11 amperes when two conductors are used to form a single-phase circuit and these are bunched or drawn into a conduit. When three or four of the same conductors are used in a conduit to form a three-phase circuit the current rating falls to 9 amperes.

Where a number of circuits are enclosed in the same conduit the following factors must be applied unless a more precise evaluation has been made by experiment or calculation.

Number of circuits	2	3	4	5	6	10	14
Factor	0.8	0.69	0.62	0.59	0.55	0.48	0.41

Example (1). Two PVC insulated conductors are drawin into a conduit and feed a single-phase load. The current rating of the conductors is 24 A under these conditions. Calculate the permissible rating if three more similar pairs are drawn into the conduit. Use the derating factors given above.

For four circuits the rating factor = 0.62.
Current rating = 0.62 × 24
= 14.9 A.

Example (2). 18 conductors in a conduit form six three-phase three-wire circuits. Each conductor has a rating of 30 A under these conditions.

To what value could the rating of each conductor be increased if (a) five circuits were removed leaving one three-phase circuit only in the conduit (b) three circuits were left in the conduit.

PERMITTED VOLT DROPS IN CABLES

I.E.E. Regulation B.23 states: 'The size of every bare conductor or cable conductor shall be such that the drop in voltage from the consumer's terminals to any point in the installation does not exceed 2.5% of the declared or nominal voltage when the conductors are carrying the full load current, but disregarding starting conditions. This requirement shall not apply to wiring fed from an extra-low-voltage secondary of a transformer. Note 1—Tables 1M to 31M show values of voltage drop per ampere per metre run of cable(s).'

A small part of Table 1M is quoted here.

For PVC insulated, non-armoured, single core conductors

Nominal cross-sectional area mm^2	Current rating for 2 cables single-phase a.c., or d.c. bunched and enclosed in conduit or trunking A	Volt drop per ampere per metre mV
1	11	40
1.5	13	27
2.5	18	16
4	24	10
6	31	6.8
10	42	4.0

Consider a circuit which is to carry a full load current of 11 A fed from a 240 V supply. Using the 1 mm^2 conductor, the volt drop per ampere per metre run is 40 mV. With full load current of 11 A, the volt drop per metre run = $11 \times 40 \times 10^{-3} = 0.44$ V.

2.5% of 240 V = 6 V. Hence the maximum length of run to keep within the regulation is 6/0.44 = 13.64 m. If the run is to be longer than this, a larger cable must be selected.

Suppose the run is to be 16 m long whilst the full load current remains unchanged. A 1.5 mm^2 cross-section cable has a volt drop of 27 mV/A/m.

Therefore the volt drop along a 16 m length with 11 A flowing
= $16 \times 11 \times 27 \times 10^{-3} = 4.75$ V.
This is less than 6 V and this cable is therefore satisfactory.

Example (3).   A single phase load of 20 A is to be fed through PVC insulated copper cables as specified in Table 1M of the I.E.E. regulations. The nominal circuit voltage is 230 V. The load is 40 m from the supply point. Determine the cross-sectional area of a suitable conductor.

2.5% of 230 V = 5.75 V. The voltage drop must not exceed this value when 20 A flows. By inspection of the table the nearest size is 4 mm^2 which has a rating of 24 A. The volt drop per ampere per metre run = 10 mV.

With 20 A flowing, over a length of 40 m the volt drop
$$= 40 \times 20 \times 10 \times 10^{-3} = 8 \text{ V}.$$
This is greater than 5.75 V so that the next size up must be tried.
Using 6 mm^2 conductors the volt drop $= 40 \times 20 \times 6.8 \times 10^{-3}$
$$= 5.44 \text{ V} \text{ which is quite satisfactory.}$$
It would be necessary to use 6 mm^2 conductors with a carrying capacity of 31 A.

Example (4). Select conductors from Table 1M for the following loads.
(a) 15 A over 10 m from a 250 V supply.
(b) 30 A over 25 m from a 200 V supply.
(c) 24 A over 22 m from a 220 V supply.

CIRCUIT DAMAGE DUE TO OVERCURRENTS

Consider a single conductor carrying a current of I_1 amperes and the resulting magnetic field as shown in *Figure 5.6*.

The magnetising force at point P is given by: $H = \dfrac{I_1}{l_1} = \dfrac{I_1}{2\pi r}$ A

where l_1 is the length of the magnetic line of force through P.
Since the magnetic flux density $B = \mu_0 H$ tesla (Strictly in vacuum, but considered true for air and solid insulators)

$$B = 4\pi \times 10^{-7} \times \frac{I_1}{2\pi r} = 2 \times 10^{-7} \times \frac{I_1}{r} \text{ tesla}$$

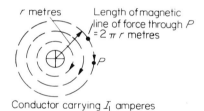

Figure 5.6

A second conductor carrying I_2 amperes running parallel to the first and through point P will suffer a force given by $F = BI_2 l_2$ newtons. The arrangement is shown in *Figure 5.7*.

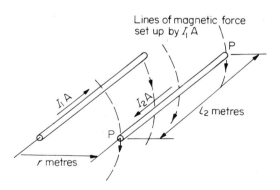

Figure 5.7

So that for $l_2 = 1$ m
$$F = 2 \times 10^{-7} \times \frac{I_1 I_2}{r} \text{ newton/metre length.}$$

Action and reaction being equal and opposite this must also be the value of the force on the first conductor. With currents in opposite directions as shown the forces are of repulsion. In single-phase a.c. and direct current circuits $I_1 = I_2$ and under normal conditions the forces involved are small. However, when a short circuit fault occurs the currents and forces can be extremely large, sufficient in fact to disrupt the cable or conductor system.

Consider two conductors 4 mm apart carrying (a) a normal current of 20 A and (b) a fault current of 2500 A.

(a) $F = 2 \times 10^{-7} \times \dfrac{20 \times 20}{4 \times 10^{-3}} = 0.02$ newtons per metre run ($l_2 = 1$)

(b) $F = 2 \times 10^{-7} \times \dfrac{2500 \times 2500}{4 \times 10^{-3}} = 312.5$ N/m run.

In three-phase circuits each conductor is situated in a magentic field produced by the currents in the other two conductors and the same effect is produced.

In addition to the forces involved the heating effect of large currents must be considerd.

Power loss $= I^2 R$ watts.

Increasing the circuit current from 20 A to 2500 A will increase the power loss by a factor $(2500/20)^2$ or 15 625 times. It is easy to envisage the effect this will have on the circuit insulation. Within a few cycles of an a.c. supply, plastic insulation will melt.

MATERIALS USED IN ELECTRONIC CIRCUITS

Because of its good conductivity and mechanical strength copper is used extensively to connect electronic components either in the form of insulated stranded wires or strip bonded to an insulating board in the form of a printed circuit. The insulation used is of polythene, polystyrene, polyurethane enamel or air where conductors can be suitably spaced.

Permanent connections between components and interconnecting copper are usually soldered. Where components or boards of components must be capable of removal from an equipment, possibly for repair, some form of sliding contact system is required. The plug and socket for connection of domestic electrical equipment is an example of such a system.

CONDUCTOR SCREENING AND COAXIAL CABLE

A conductor carrying current sets up a magnetic field around it, the strength of which is proportional to that current. This magnetic field may affect other circuits in its vicinity and to prevent this circuit screening is necessary.

Where a component is to be screened from a steady magnetic field it is surrounded with a material with low magnetic reluctance. In *Figure 5.8* the component to be screened is marked C. Without the screen the magnetic field will affect the component. With the screen in position the magnetic lines of force take the low reluctance path which is through the nickel-iron, so leaving the component in a position of zero flux.

Where alternating current is involved the conductor carrying the current, and hence producing the alternating magnetic flux, is surrounded by the screen. The arrangement is in effect a transformer, the conductor being the primary and the screen a short circuited secondary.

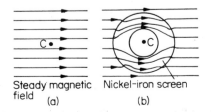

Figure 5.8

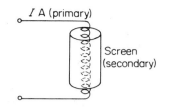

Figure 5.9

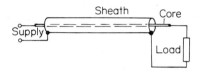

Figure 5.10

Figure 5.11

The alternating flux induces a voltage in the screen and a current flows which sets up an opposing magnetic flux. The two fluxes are very nearly equal so that the magnetic field outside the screen is virtually zero.

The magnetic field around a cable feeding a circuit is eliminated by using the coaxial construction. This is particularly important at very high frequencies when non-screened cables can cause considerable interference with other circuits.

Current is supplied to its load along the core of the cable and is returned to the supply through the sheath which completely surrounds the core. Since the go and return currents are in opposite directions, the magnetic fields produced are in opposite directions and completely cancel each other outside the sheath.

The cable may be formed using a solid core with a copper tube for the sheath with solid insulation (M I.C.S.) or by using a solid or stranded core with a braided outer sheath for flexibility when the insulation is generally polythene.

For many applications air is the best insulation but air alone is not practicable because the spacing between the core and sheath has to be maintained. A compromise between solid insulation and air is achieved by using either plastic foam insulation containing a substantial amount of air in the form of bubbles or by using a polythene spiral wound on the inner core, when again the insulation is substantially air.

For extra-high-voltage power the concentric construction is used with paper insulation (see *Figure 1.4*).

PRINTED CIRCUITS

The electrical connections between circuit components must have low resistance and this is often achieved by using insulated copper wire or bare copper or aluminium strip. The working components are then generally individually soldered to the conductors. In equipments with many components there are several problems to overcome. One is that of the dry joint, when the soldering looks sound but in fact has not made good contact and the joint has a high resistance. Another problem with individual hand wiring is that it is difficult to reproduce a circuit accurately in successive equipments. There may be errors in connections and slightly different lengths and routing for wiring taken. Circuit response is thereby affected. Fault finding can be very difficult and it may be necessary to unsolder connections to do tests.

Mounting components on an insulating board to a fixed pattern is one solution to these problems. This is especially true now that valves have been largely replaced by transistors and integrated circuits in which the heat generated is relatively small. The fixed pattern ensures repeatable values of circuit resistance and capacitance.

The board must provide mechanical support for the components and the means for their interconnection. The layout of components is much clearer than with individually wired equipment and the values of the components are often marked on the board. Servicing of the equipment is simpler since faulty boards may be replaced to get the equipment working and the components on the faulty boards replaced at a central workshop.

Modern rigid boards are made of phenolic or epoxy resins reinforced with woven glass fabric or paper. They are available up to about one metre square. Flexible boards are available and these are made from polyester film or for high temperature work from PTFE or polypropylene possibly reinforced with glass fibre.

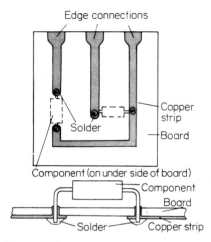

Figure 5.12

Printed circuit boards employ copper foil conductors securely cemented to one of the above laminates. Holes are drilled through the foil and board and the component tails are fixed by soldering. *Figure 5.12* shows a very simple board with two components.

To construct a printed circuit board it is necessary to draw a master circuit diagram. Larger areas of foil have to be allowed for parts of the circuit which will carry the heaviest currents. The master circuit can be drawn directly on to the board in the case of a prototype or it can be printed on using a photographic process when large numbers of identical circuits are required.

Two processes are available for the production of the final circuit.

1. The subtractive method.   A board is obtained which has foil completely covering one side. The diagram is drawn or printed on the foil using acid-resisting paint and the board is dipped into acid which dissolves the copper which is not protected. Only the required circuit is left on the board and the necessary holes can be drilled for circuit assembly.

2. The additive method.   Insulating board is used with no foil covering. The circuit is drawn or printed on the board using conducting paint. Copper is deposited from a plating bath over the treated area. Other additive methods involve the use of metallic powder and heat or foil strips and mechanical force.

The current carrying capacity of the foil is determined largely by the permissible temperature rise. Charts showing current carrying capacities and cross-sectional areas for different increases in temperature are consulted when circuits are being designed. Foils vary in thickness from 0.035 mm to 0.106 mm and in width from 0.25 mm up to several millimetres.

As an example a strip 0.07 mm thick and 0.76 mm wide can carry 3.5 A allowing a 40°C rise in temperature but only 2 A allowing a 10°C rise, both from 15°C.

The cross-sectional area of this strip = 0.07 × 0.76 = 0.053 mm²

Allowing a 40°C rise, the permissible current density = $\dfrac{3.5}{0.053}$ = 65.8 A/mm²

Allowing a 10°C rise, the permissible current density = $\dfrac{2}{0.053}$ = 37.6 A/mm²

Increasing the strip width gives a less than proportional increase in current carrying capacity so that current for other strips cannot be deduced by proportion.

The current densities obtained in these very small conductors are considerably greater than those which can be achieved with mains cables since the insulation is very much thinner and the surface area available for cooling is greater per given volume.

To connect the board into the main circuit either plug and socket or edge connectors are used.

Figure 5.13 shows the form of the edge connector. Where large boards are used there may be over 100 connections to make so that considerable force may be necessary to insert a multi-pin plug into its socket or the board into its edge connector. When using such force it is difficult to determine whether a correct match of the contacts has been made or whether in fact some of them are being badly deformed or crumpled up.

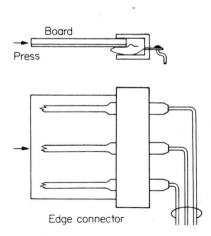

Figure 5.13

There are several patent ways of achieving low insertion pressure followed up by clamping of the board in position. The situation is eased considerably by the choice of contact materials. For example the pressure needed between gold or platinum contacts to obtain very low contact resistance is only 0.03 times that necessary to obtain the same resistance between brass contacts. The value for nickel and silver is 0.3 times that for brass.

Gold is generally used for low voltage connections of this type since in addition to the low contact pressure required, it does not oxidise or corrode. Even a very thin oxide film can cause a virtual open circuit unless the voltage employed is large enough to break it down. The gold, generally less than 50×10^{-6} m thick, is plated over silver or nickel, the latter combination being especially good in hostile environments.

Interconnection between boards is achieved using insulated wires which are grouped into multi-core cables or merely bunched and neatly carried round the chassis of the equipment. Great care has to be taken to ensure that the capacitance between cores in such close proximity does not affect the circuit performance, that all the connections are sound and that any one circuit does not affect any other as already discussed in the section 'Conductor screening and coaxial cable'.

MULTI-CORE CABLES

Multi-core cables are used where many circuits are required to run over the same route between a pair of terminal points, for example for the control of equipment, metering and indication circuits and transmission of condition information such as temperature, pressure, position and power consumption. Power to the equipment concerned is usually fed through a separate cable to minimise the risk of damage to low voltage equipment should a fault occur.

The cables comprise the required number of cores with fibre fillers if necessary to make up a circular cross section, with a belt of insulation and armouring if required. Each core is numbered or otherwise identified along its length.

The heavy current types in common use are:

1. Stranded copper conductors with paper or fabric insulation with overall paper wrapping, lead covering, steel wire armoured with an anti-corrosion sheath. The cable is terminated at each end in a compound filled box.

2. Stranded copper conductors with PVC insulation, PVC belt, armoured with PVC anti-corrosion sheath overall.

Each conductor is terminated in a soldered or crimped lug which is bolted to a terminal block from which connections to equipment are made.

Lighter types comprise the required number of single strands of fine copper wire insulated with PVC and covered with a belt of PVC. End connections may be crimped as before or bound and soldered to terminal posts.

Multi-core cables for telecommunications have pairs of fine insulated wires twisted together along their length to form circuits, two pairs then being twisted together. Many such groups of four conductors make up a complete cable.

For the interconnection of printed circuit boards, flexible flat cables may be used. The spacing between the conductors is the same as that for the contacts in the edge connectors. Cables with 50 or even more conductors are available.

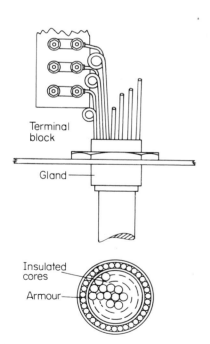

Terminal block

Gland

Insulated cores

Armour

Figure 5.14 PVC insulated multi-core cable

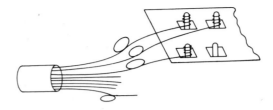

Figure 5.15

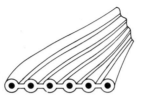

Figure 5.16

HEAT SINKS

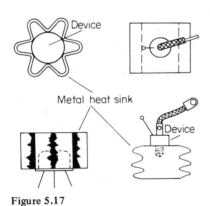

Figure 5.17

The characteristics of semiconducting devices such as rectifiers and transistors change considerably with a rise in temperature so much so that the circuit in which they are connected may cease to operate in the manner intended. For example, a rectifier will become conducting in both directions if it is made hot enough. Often a device is permanently damaged and the original characteristics are not re-established by cooling. Where excessive currents have been drawn due to these changes, other components in the circuit may have been damaged.

Insulation deteriorates at high temperature and the printed circuit board itself or connecting wires can suffer damage. With plug-in components the contacts may suffer since contact springs can become soft and ineffective so introducing a high resistance and more heat into the circuit.

Heat sinks are metal clip or screw-on additions to a device which effectively increase the surface area and so aid cooling.

Two types of heat sink are shown in *Figure 5.17*. These may be fan cooled in extreme cases.

THE INSULATION OF OVERHEAD LINES

The material most commonly used for overhead line insulators is porcelain. The insulators are shaped from the raw material: a mixture of clay, finely ground feldspar and silica in water, is dried, dipped in liquid glaze and then fired at very high temperature. The glaze forms a glass-like coating providing a surface to which dirt cannot readily stick It also improves the strength of the insulator so that fracture is more difficult.

One disadvantage of porcelain is that when the glaze is chipped by a power arc or by missiles projected by vandals, water can soak into the body of the insulator. This causes it to become conducting and an electrical discharge takes place which trips out the line while the heat produced dries out the insulator. Such a fault is immensely difficult to find, the line tripping out each time there is rain.

Glass is an alternative material. Toughened glass insulators have a higher breakdown strength under electrical stress than porcelain and if damaged shatter completely rather like the windscreen of a motor vehicle when it is hit by a stone. Missing insulators leave gaps which are easily spotted during one of the regular inspections which are made. The construction is such that although the insulator shatters the line cannot fall down.

The glass is toughened by heating it to the softening point and then cooling it fairly rapidly. This causes the surface to become hard whilst the centre is still plastic. As the centre cools and sets it tries to contract so pulling in on the outer layers. Before the insulator can fracture, these internal stresses have to be overcome and the force required can be six times that required to break ordinary glass.

Suspension insulators are made of porcelain or glass whilst pin insulators are almost invariably made of porcelain (see *Figure 1.3*).

Insulators must withstand mechanical and electrical stresses. Heavy lines must be held off the ground whilst the electrical potential of the line is considerably above that of earth. Even in wet weather the insulator must function and for this reason it has sheds or skirts to keep at least part of the surface dry in almost any weather conditions.

Figure 5.18 shows a pin type insulator supported at the bottom with the line at the top. The steel pin is at earth potential. An electrical breakdown or flashover can occur in one of three ways:

1. A dry flashover can occur. This means that an arc will form round the insulator from line to pin along the route 1 in *Figure 5.18*.

2. A wet flashover can occur. When the top surfaces of the insulator are wetted by rain they become conducting so that the sheds are so disposed to keep the undersides dry. A flashover can occur along the route 2 in *Figure 5.18*.

3. Current can leak from the line over the surface of the insulator along the route 3 in *Figure 5.18*.

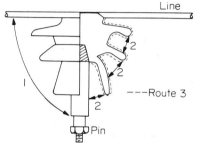

1 Dry flashover distance
2 Wet flashover distance
(exposed surfaces wet & conducting)
3 Dry leakage distance

Figure 5.18

Atmospheric pollution causes dirt to accumulate on the surface of the insulator making it partly conducting and a flashover more likely. Fog wets the insulator overall causing surface leakage to occur especially when smoke and sulphur oxides are also present. Sea spray carried by the wind also has the same effect.

PROBLEMS FOR SECTION 5

(5) Although silver is the best electrical conductor it is not used in cable manufacture. Why is this?

(6) How is the low tensile strength of aluminium compensated for in the construction of overhead lines?

(7) Hard drawn copper is not very flexible. How is flexibility built into cables and lines with copper cores?

(8) Why would PVC insulation not be used in very hot situations?

(9) What is the power loss per metre run of a copper conductor 2.5 mm^2 in cross-sectional area when carrying 15 A?

(10) A single phase circuit in a conduit has a current-carrying capacity of 15 A. To what value would the permissible current rating fall if two additional identical circuits were drawn into the conduit?

(11) A single-phase circuit carrying 50 A has a power loss of 1 W per metre run due to the conductor heating. There is a force

between the conductors of 0.1 N per metre run due to the magnetic effect.

Calculate the power loss and force between the conductors when a short circuit occurs and the current rises to 5000 A.

(12) Why is it necessary to use coaxial cables when interconnecting equipments operating at high frequencies?

(13) What advantages are there to using printed circuit boards in a television receiver as compared with an individually wired chassis?

(14) Describe two processes which are available for the manufacture of printed circuit boards.

(15) A printed circuit board has one circuit formed from foil 0.035 mm thick and 0.5 mm wide. The permitted current density for a $40°C$ temperature rise is 50 A/mm^2. Calculate the maximum permissible current in this circuit.

(16) Calculate the current densities in the following printed circuit board conductors.

(a)   0.07 mm × 1 mm carrying 5 A.

(b)   0.106 mm × 2 mm carrying 4 A.

(17) What is the function of a heat sink?

(18) Name the two principal insulating materials used in overhead line construction.

(19) What is the difference between a pin insulator and a suspension insulator?

(20) Why is flashover on an overhead line insulator more likely to occur on a rainy day than on a dry day? What features are built into the insulator to minimise the occurrence of flashover?

6 Single phase transformers

Aims: At the end of this section you should be able to:
Describe the open and short circuit tests on power transformers.
Draw the on-load phasor diagram using data from the above tests.
Calculate transformer efficiency and determine the load at which maximum efficiency occurs.
Describe the construction of the auto-transformer and list the advantages and disadvantages of the connection.
Explain the function of instrument transformers.

Faraday discovered that whenever a change in magnetic flux is associated with a coil of wire a voltage is induced in that coil. The value of the induced e.m.f. is proportional to the number of turns and to the rate of change of magnetic flux in webers per second.

$$e = N \frac{d\Phi}{dt} \text{ volts}$$

Alternating voltages of any desired value may be obtained by using the transformer which employs this principle. Voltages need to be changed between the points of generation and the consumer several times in order to arrive at the most economical levels for transmission and distribution. Generation is carried out at voltages between 11 kV and 25 kV whilst major transmission voltages are 275 kV and 400 kV. Domestic consumers are supplied at about 240 V.

PRINCIPLE OF ACTION OF THE TRANSFORMER

Figure 6.1 shows the general arrangement of a transformer with the secondary open circuited. There are two coils, generally known as the primary and secondary, wound on an iron core. The iron core is made up of laminations which are about 0.3 mm thick. These have been rolled to the correct thickness, acid cleaned, polished and varnished or anodised on one side, and then made up into the correct core form.

When an alternating voltage V_p is applied to the primary coil a small magnetising current flows which sets up a magnetic flux in the iron core. This alternating flux links with both the primary and secondary coils and with the iron of the core inducing voltages in each. The voltage E_p induced in the primary coil opposes the applied voltage V_p according to Lenz's law. The difference between V_p and E_p is very small. The voltage induced in the iron core causes eddy currents to flow so giving rise to the production of heat. Dividing the core into well insulated laminations increases the resistance so minimising these currents and the associated loss. The eddy current and hysteresis losses due to alternating magnetisation must be provided by the power source.

Finally the voltage E_s induced in the secondary winding is used to supply the load.

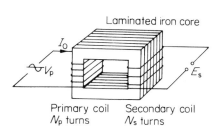

Laminated iron core

Primary coil N_p turns   Secondary coil N_s turns

Figure 6.1

E.M.F. EQUATION   Let the maximum value of core flux be Φ_m webers and the frequency f hertz.

The time taken for the flux to change from $+\Phi_m$ to $-\Phi_m$ is $\dfrac{\tau}{2}$ or $\dfrac{1}{2f}$ seconds.

Since $e = N \dfrac{d\Phi}{dt}$ volts

the average e.m.f. induced in the primary winding $= N_p \times 2\Phi_m \div \dfrac{1}{2f}$

$= 4N_p\Phi_m f$ volts.

Where N_p = number of turns on the primary winding.

For a sine wave, the r.m.s. value is 1.11 times the average value. Therefore $E_p = 4.44 \, N_p\Phi_m f$ volts and if all the magnetic flux set up by the primary winding links with the secondary, $E_s = 4.44 \, N_s\Phi_m f$ volts, where N_s = number of turns on the secondary.

Therefore $\dfrac{E_s}{E_p} = \dfrac{N_s}{N_p}$

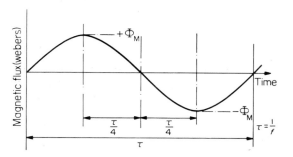

Figure 6.2

THE NO-LOAD PHASOR   Commencing with the primary applied voltage V_p.
DIAGRAM   The core losses must be provided by a current in phase with the supply voltage and this is shown as I_{H+E} in *Figure 6.3a*.

Power loss in the core $= V_p \times I_{H+E}$ watts.

Consider the primary coil to have no resistance and therefore to be a perfect inductor. The magnetising current flowing in it lags the applied voltage by exactly 90°. This current I_m sets up the flux in the core. *Figure 6.3b* shows the associated phasors. Both the loss current and the magnetising current flow simultaneously and the phasor sum of the two currents is I_o, the standing no-load current of the transformer. The phasor addition is shown in *Figure 6.3c*.

Since E_p is very nearly equal and opposite to V_p it is added to the phasor diagram as shown in *Figure 6.4*. E_p and E_s are both produced by the same flux and are therefore in phase.

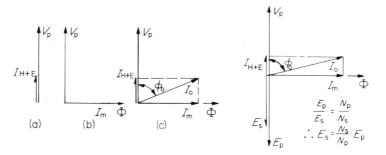

Figure 6.3   Figure 6.4

THE OPEN CIRCUIT TEST

The normal rated voltage is applied to one of the windings of the transformer while the other winding is connected to a high resistance voltmeter. The input voltage, power and current are measured using suitable range instruments connected as shown in *Figure 6.5*. Since the second winding is delivering virtually no current and the current in the winding being fed is very small, the total input power to the transformer as indicated on the wattmeter may be considered to be the core loss.

Using the results from this test the no-load phasor diagram can be constructed.

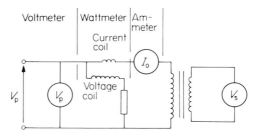

Figure 6.5   Open circuit test

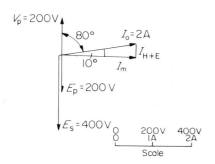

Figure 6.6

Example (1).   An open circuit test on a 200 V:400 V single phase transformer gave the following results with the 200 V winding connected to a 200 V supply:

Input current = 2 A.
Input power = 69.44 W.

Calculate: (a) the no-load power factor and phase angle.
             (b) the value of the iron loss current I_{H+E}
             (c) the value of the magnetising current I_m.
Draw a phasor diagram showing these quantities together with E_p and E_s.

$$\text{(a) Power factor} = \frac{W}{VI} = \frac{69.44}{200 \times 2} = 0.174 = \cos \phi$$

Therefore $\phi = 80°$

$$\text{(b)} \ \frac{I_{H+E}}{2} = \sin 10° \quad \text{Therefore } I_{H+E} = 2 \times 0.174 = 0.348 \text{ A.}$$

$$\text{(c)} \ \frac{I_m}{2} = \cos 10° \qquad I_m = 2 \times 0.9848 = 1.969 \text{ A.}$$

BALANCING CURRENT

Consider now the effect of connecting the secondary terminals to a load so causing a current I_s to flow in the secondary winding. I_s amperes flowing in N_s turns is a magnetomotive force which creates a flux which modifies the flux in the core set up by the magnetising current flowing in the primary winding.

Since $E_p = 4.44 \, N_p \Phi_m f$ volts, any change in core flux changes the value of E_p. The applied voltage V_p will differ from E_p and a current I_s' will flow in the primary winding such that the extra magnetomotive force $I_s' N_p = I_s N_s$

Transposing $I_s' = I_s \dfrac{N_s}{N_p}$

I_s', the secondary current referred into the primary, is known as the balancing current.

The balancing current flows at the same time as I_o so that in the phasor diagram for the on-load condition these two currents must be added.

In *Figure 6.7*, I_s' is drawn to balance I_s and therefore lags V_p by the same phase angle as I_s lags E_s. Adding I_s' phasorially to I_o gives I_p, the total primary current. I_p lags V_p by an angle ϕ_p which is the operating power factor of the transformer for that particular load.

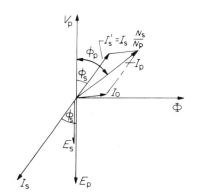

Figure 6.7

Example (2).   A single phase transformer with ratio 500 V:100 V has a core loss of 200 W at full rated voltage. The magnetising current is 2 A in the 500 V winding.

The transformer delivers a current of 50 A from its 100 V winding at a power factor of 0.75 lagging.

Determine: (a) the iron loss component of current.
(b) the no-load current of the transformer.
(c) the balancing current in the primary winding.
(d) the total primary current.
(e) the power factor of the primary input at this load.

(a) From *Figure 6.8a* it may be seen that $I_o \cos \phi = I_{H+E}$.
But power input on no load = $V_p I_o \cos \phi$ watts.

Therefore $500 \times I_{H+E} = 200 \quad I_{H+E} = \dfrac{200}{500} = 0.4$ A

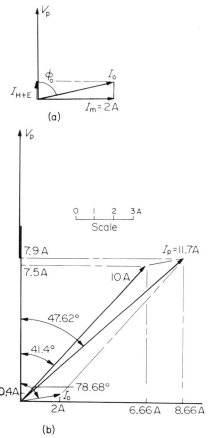

(a)

O 1 2 3A
Scale

(b)

Figure 6.8

(b) $I_o = \sqrt{(I_{H+E}{}^2 + I_m^2)}$ (Pythagoras) $= \sqrt{(0.4^2 + 2^2)} = 2.039$ A.

$$\cos\phi_o = \frac{I_{H+E}}{I_o} = \frac{0.4}{2.039} = 0.196. \quad \text{Therefore } \phi_o = 78.68°.$$

(c) $I_s' = 50 \times \dfrac{N_s}{N_p} = 50 \times \dfrac{E_s}{E_p} = 50 \times \dfrac{100}{500} = 10$ A.

(d) The load power factor $= 0.75$ so that I_s' lags on V_p by $41.4°$.

These quantities are drawn to scale in *Figure 6.8b*.
The value of I_p and the overall phase angle may be scaled from the diagram or calculated by resolving I_s' into vertical and horizontal components.

Vertical component of $I_s' = 10\cos 41.4° = 7.5$ A.
Horizontal component of $I_s' = 10\sin 41.4° = 6.6$ A.
Adding vertical components: $7.5 + 0.4 = 7.9$ A.
Adding horizontal components: $6.6 + 2 = 8.6$ A.
By Pythagoras' theorem: $I_p = \sqrt{(7.9^2 + 8.6^2)} = 11.7$ A.

$$\cos\phi_p = \frac{7.9}{11.7} = 0.675 \quad \text{Therefore } \phi_p = 47.6°.$$

Example (3). The magnetising current in the 400 V primary winding of a 400 V:100 V 50 Hz, single phase transformer is 1 A whilst the total no-load current is 1.3 A. The primary winding consists of 600 turns.
Calculate: (a) the loss component of the no-load current (b) the core loss in watts (c) the no-load power factor (d) the number of turns on the secondary (e) the maximum value of the core flux in webers.

Example (4). A transformer has a ratio 3 300 V:240 V. It supplies a current of 550 A at 240 V. Calculate the value of the balancing current in the 3 300 V winding.

Example (5). A 440 V:110 V single phase transformer draws a magnetising current of 1 A and an iron loss current of 0.25 A, both in the 440 V winding. The secondary load at 110 V is 10.7 A at a power factor of 0.85 lagging.
Calculate: (a) the no-load current of the transformer (b) the balancing current in the primary winding (c) the total primary current (d) the phase angle between the primary current and the supply voltage.

RATING   A transformer has a name plate rating which in effect informs us how much current the transformer windings can carry without overheating. The rating is quoted in volt-amperes (VA), at the full rated voltage. For example, a single phase transformer with a rating of 10 000 VA and a ratio of 500 V:100 V can carry 10 000/500 = 20 A in its 500 V winding and 10 000/100 = 100 A in its 100 V winding. If the voltage is lower than that specified on the name plate for any reason, the current may not be increased to give the original value of VA. Such an increase in current would increase the power loss in the windings and the transformer would overheat.

Example (6).   Determine the rated current in each winding of a 1 100 V:240 V single phase transformer with a rating of 25 kVA.

1100 V windings: $I = \dfrac{25\,000}{1100} = 22.72$ A

240 V winding: $I = \dfrac{25\,000}{240} = 104.2$ A.

SHORT CIRCUIT TEST

The short circuit test is used to determine the power loss in the transformer windings.

Winding power loss $= (I_s^1)^2 R_p + I_s^2 R_s$ watts, where R_p and R_s are the resistance of the primary and secondary windings respectively.

One winding of the transformer is short circuited through an ammeter whilst the other winding is fed from a variable voltage supply through a wattmeter and ammeter as shown in *Figure 6.9*. The input voltage is raised in increments from near zero to a value at which full load currents are circulating in both windings. The input power, voltage and currents are recorded at each step.

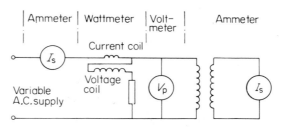

Figure 6.9   Short circuit test

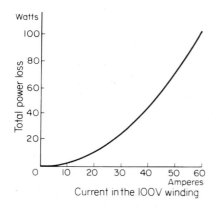

Figure 6.10

Usually about 10% of the rated voltage is sufficient to cause full-load currents to flow, and from the e.m.f. equation it can be seen that this will require only a low value of core flux and therefore of flux density. Since the core losses are very nearly proportional to the square of the maximum value of flux density, these will be very small. At 10% of normal voltage and flux density the core losses will be $(10\%)^2 = 1\%$ of those measured in the open circuit test (closely).

The power losses in this test are therefore considered to be due to winding resistances only.

The losses are proportional to the square of the current, and a graph of power loss in both windings as indicated by the wattmeter in *Figure 6.9*, for various load currents in a 200 V:100 V transformer, is shown in *Figure 6.10*.

EFFICIENCY The two losses in a transformer are the core or iron loss which is measured in the open circuit test, and the conduction or copper losses which are determined using the short circuit test.

$$\text{Efficiency} = \frac{\text{Output power}}{\text{Input power}} = \frac{\text{Output power}}{\text{Output power} + \text{losses}}$$

$$= \frac{\text{Output VA} \times \text{power factor}}{\text{Output VA} \times \text{power factor} + \text{core loss} + \text{winding losses}}$$

Example (7). A single phase transformer rated at 15 kVA has a core loss of 400 W and a winding loss of 540 W when full load current flows in the windings. Calculate the efficiency of the transformer for each of the following loads.
(a) Full load at unity power factor
(b) Full load at 0.6 power factor lagging
(c) Half full load at unity power factor.

(a) Efficiency = $\dfrac{15\,000 \times 1}{15\,000 \times 1 + 400 + 540}$ = 0.94

(b) Efficiency = $\dfrac{15\,000 \times 0.6}{15\,000 \times 0.6 + 400 + 540}$ = 0.905

Note that although the output *power* is different at 0.6 power factor, the number of kVA and hence current and winding losses are the same as in (a).

(c) At half full load the winding losses will be reduced by a factor $(1/2)^2$ since the losses are proportional to I^2

Winding loss on half load = $\frac{1}{4} \times 540 = 135$ W
The core loss does not change
The load is $\dfrac{15\,000}{2}$ = 7500 VA

Efficiency = $\dfrac{7500 \times 1}{7500 \times 1 + 400 + 135}$ = 0.93

Example (8). A 480 V:240 V, 20 kVA single-phase transformer has a core loss of 200 W and a copper loss of 45 W when one half full load current flows in the windings. Calculate the efficiency of the transformer when delivering:
(a) 10 kVA at 0.8 power factor lagging
(b) 20 kVA at unity power factor
(c) 15 kVA at 0.75 power factor lagging.

CONDITION FOR MAXIMUM EFFICIENCY The maximum efficiency of a transformer occurs at a load at which the winding copper loss is equal to the core loss. *Figure 6.11* shows the core and winding losses of a transformer and a typical efficiency cure. The maximum efficiency generally lies between 0.95 and 0.99 depending upon type and rating.

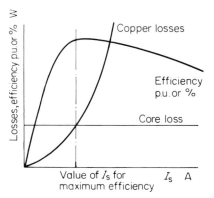

Figure 6.11

THE AUTO-TRANSFORMER

The auto-transformer has only one winding. Part of this winding is common to both primary and secondary which are therefore both electrically and magnetically linked.

Figure 6.12 shows the possible arrangements of coils and core. The inputs and outputs are reversible providing for voltage increase or decrease. Considering the transformers to be ideal, i.e. ignoring all losses, the simplified circuits shown in *Figure 6.13* may be drawn.

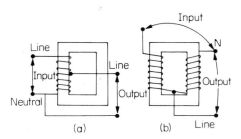

Figure 6.12

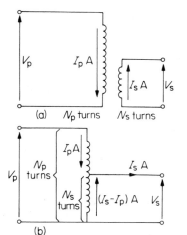

Figure 6.13

Figure 6.13a shows the current directions in the primary and secondary of a double wound transformer. When the secondary is in fact part of the primary, the current in the secondary section becomes $(I_s - I_p)$ as shown in *Figure 6.13b* and the cross-sectional area of this section may be reduced so saving copper.

Example (9). A single phase auto-transformer has a ratio 500 V : 400 V and supplies a load of 30 kVA at 400 V. Calculate the value of current in each section of the winding. Assume ideal operation.

30 kVA at 500 V requires $\dfrac{30\,000}{500} = 60$ A

30 kVA at 400 V requires $\dfrac{30\,000}{400} = 75$ A

Current in the secondary section of winding = 75 − 60 = 15 A.

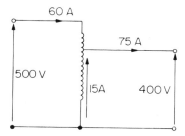

Figure 6.14

ADVANTAGES OF THE AUTO OVER THE DOUBLE-WOUND TRANSFORMER

1. *Less copper is required.* The volume of copper in a winding is proportional to the number of turns and to the cross-sectional area of the wire used, which is in turn proportional to the current to be carried. Therefore   Volume of copper $\propto NI$

For a double wound transformer

Volume of copper $\propto N_p I_p + I_s N_s$ and assuming ideal operation $N_p I_p = I_s N_s$ Therefore   Volume of copper $\propto 2 N_p I_p$

For an auto-transformer (see *Figure 6.13*)

Volume of copper $\propto N_s(I_s - I_p) + (N_p - N_s)I_p$

$$\propto N_s I_s + N_p I_p - I_p N_s - I_p N_s$$

$$\propto N_s I_s + N_p I_p - 2 I_p N_s \text{ and since } N_s I_s = N_p I_p$$

$$\propto 2 N_p I_p - 2 I_p N_s$$

$$\frac{\text{Volume of copper in the auto-transformer}}{\text{Volume of copper in the double-wound transformer}} = \frac{2 N_p I_p - 2 I_p N_s}{2 N_p I_p}$$

$$= 1 - \frac{N_s}{N_p}$$

Transposing

Volume of copper in the auto-transformer

$$= \left(1 - \frac{N_s}{N_p}\right) \times \text{volume of copper in a double-wound transformer.}$$

When N_s approaches N_p (V_s approaches V_p) the saving in copper is greatest.

Example (10). Compare the volume of copper in a single-phase auto-transformer with that in a double-wound transformer for ratios (a) 400 V:300 V (b) 400 V:50 V.

(a) Volume of copper in the auto-transformer = $\left(1 - \dfrac{300}{400}\right) \times$ volume in the double-wound transformer = 0.25 times

(b) Volume of copper in the auto-transformer = $\left(1 - \dfrac{50}{400}\right) =$ 0.875 times that in the double-wound transformer.

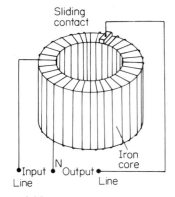

Figure 6.15

2. *The weight and volume of the auto-transformer is less.*

3. *The auto-transformer has a higher efficiency* and suffers less voltage variation with changing load due to the better magnetic linkage between the primary and secondary sections of the winding.

Auto-transformers are used to interconnect the 400 kV, 275 kV and 132 kV sections of the British grid system. When transporting very large high voltage transformers by road the weight has to be carefully considered and quite often, before major power system construction is commenced, roads and bridges have to be specially reinforced to carry such loads.

It is also worth noting that an increase in efficiency of only 0.5% means a reduction in losses of 2500 kW in a 500 000 kW transformer.

Auto-transformers are also used to reduce the voltage supplied to induction motors and to increase the voltage supplied to discharge lamps during their starting periods.

4. *A continuously variable output voltage is obtainable.* Using the arrangement shown in *Figure 6.15* a continuously variable output voltage may be obtained. This is used, for example, in laboratories and the variable voltage required for the short circuit test shown in *Figure 6.9* could be obtained in this manner.

DISADVANTAGES OF THE AUTO-TRANSFORMER

1. *Since the neutral connection is common to both primary and secondary*, earthing the primary automatically earths the secondary. Double-wound transformers are sometimes used to isolate equipment from earth.

2. *If the secondary suffers a short circuit fault*, the current which flows will be very much larger than in the double-wound transformer due to the better magnetic linkage. There is more risk of damage to the transformer and circuit due to heating and the mechanical forces set up between current-carrying conductors.

3. *A break in the secondary section of the winding* stops the transformer action and the full primary voltage will be applied to the secondary circuit.

CURRENT TRANSFORMERS

The current required to give full-scale deflection of a d.c. moving-coil ammeter is very small being typically only a few milliamperes. When large currents are to be measured a shunt or bypass resistor is used in conjunction with the meter. Alternatively a moving-iron meter may be used.

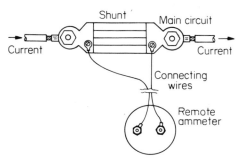

Figure 6.16

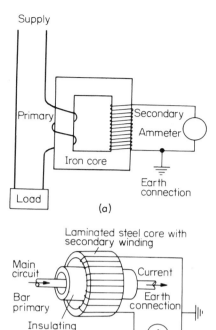

(a)

Laminated steel core with secondary winding

(b)

Figure 6.17

In either case the meter coil is at the potential of the circuit in which the current is being measured. With very large currents the size of the conductor and meter terminals will be large and the internal wiring of control and metering panels made very unwieldy. An alternative using an external shunt is shown in *Figure 6.16*. This involves individual calibration of each such meter with its shunt to allow for the resistance of the connecting wires.

Where alternating currents are involved a shunt cannot be used since the proportion of the current which flows in the meter will depend on its impedance, which varies with frequency. A small change in frequency would upset the calibration of the meter.

These problems are overcome by the use of current transformers which isolate the meter from the main circuit and allow the use of a standard range of meters giving full-scale deflections with 1, 2 or 5 A irrespective of the value of current in the main circuit. Two types of current transformer are shown in *Figure 6.17*.

Figure 6.17a has a wound primary while *Figure 6.17b* has a bar primary.

Figure 6.18 shows typical terminal markings for current transformers.

The primary of the current transformer is connected in series with the load on the circuit, replacing an ammeter, and has an extremely small voltage drop. The core flux is therefore small. The value of the primary current is determined entirely by the load in the main circuit and not by the load on its own secondary which is typically between 2.5 VA and 30 VA.

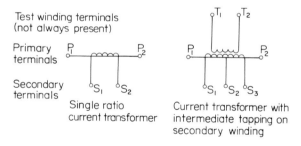

Figure 6.18

As in the power transformer $I_s' N_p = I_s N_s$

Transposing $I_s = I_s' \dfrac{N_p}{N_s}$

Since the core flux is small, the balancing current I_s' may be considered to be equal to I_p.

$$I_s = I_p \frac{N_p}{N_s} \quad \text{(Closely)}$$

The magnetic fluxes set up by the m.m.f.s $I_p N_p$ and $I_s N_s$ may individually be quite large but are very nearly equal and are in opposite directions.

Example (11).   A current transformer has a primary winding of 2 turns and a secondary winding of 100 turns. The secondary winding is connected to an ammeter with a resistance of 0.2 Ω. The resistance of the secondary of the current transformer is 0.3 Ω. The value of current in the primary winding is 250 A.
Calculate: (a) the value of current in the C.T. secondary
             (b) the potential difference across the ammeter terminals
             (c) the total load in VA on the C.T. secondary.

(a) $I_s = I_p \dfrac{N_p}{N_s} = 250 \times \dfrac{2}{100} = 5$ A.

(b) Potential difference across the ammeter terminals
$\qquad = I_s \times$ resistance of the meter
$\qquad = 5 \times 0.2$
$\qquad = 1$ V.

(c) Total resistance of the secondary circuit = 0.2 + 0.3 = 0.5 Ω.
$\qquad$ Since 5 A is flowing, the total induced e.m.f. $= 5 \times 0.5$
$\qquad\qquad\qquad\qquad\qquad\qquad\qquad\qquad\qquad\quad = 2.5$ V.

$\quad$ Total VA $= 2.5 \times 5$
$\qquad\qquad\quad = 12.5$ VA.

Example (12).   A current transformer has a bar primary (1 turn). The secondary is connected to an ammeter which indicates 4.3 A when the current in the main circuit is 344 A. Determine the number of turns on the C.T. secondary.

The secondary of a C.T. must never be open circuited whilst the primary is carrying current since under these conditions $I_s N_s$ will be zero. In a power transformer this would cause I_s' to fall to zero so that the core flux would remain at its normal value. In the C.T. the primary current is determined by the load in the main circuit and therefore does not fall when the secondary C.T. circuit is disconnected.

The flux set up by the primary m.m.f. will be unopposed and will link with the secondary inducing a large voltage in it. This may be a danger to life and to the insulation within the C.T. The large flux will also cause an increase in the hysteresis and eddy current losses in the core with subsequent heating and further damage to the insulation. The C.T. may well be ruined.

The secondary of the C.T. is earthed to prevent its potential rising above that of earth due to the capacitance between the secondary and the high voltage primary. Also in the event of an insulation failure between primary and secondary, the earth connection would allow fault current to flow which should cause the circuit to be isolated. (Electricity Supply Regulation 20).

Typical C.T. ratings are given in B.S. 3938 and are 10, 15, 20, 30, 50 or 75 A in the primary with 1, 2 or 5 A in the secondary. Current transformers with bar primaries are made for circuits carrying several thousands of amperes however.

VOLTAGE OR POTENTIAL TRANSFORMERS

A d.c. ammeter is converted into a voltmeter by the addition of a series resistor or multiplier which limits the current at full rated voltage to that required to give full scale deflection of the movement. Up to about 1000 V this arrangement is quite satisfactory. Insulating the meter movement, the terminals and the multiplier from the case and the panel in which the meter is situated presents no special problems.

Above 1000 V hazards begin to present themselves. The cables to the meter may be long and are vulnerable to damage. Insulation becomes difficult and increasingly expensive as voltages rise.

Where alternating voltages are to be measured, the voltage transformer is used to reduce the voltage at the meter to 63.5 V or 110 V typically.

The voltage transformer is essentially a power transformer designed to minimise the core loss. Great care is taken to obtain maximum flux linkage between the coils and the winding resistance is made very small by using conductors with a larger cross-sectional area than in a power transformer of similar rating. The secondary is used for measuring purposes only so that the current is small. The internal volt drops may generally be ignored so that

$$\frac{V_p}{V_s} = \frac{N_p}{N_s} \text{ (Closely)}$$

Figure 6.19 shows a voltage transformer connected to one phase of an 11 kV system. The transformer has a ratio of 100:1. The dial of the voltmeter is marked to indicate the voltage on the high voltage side allowing for the transformer ratio.

Where voltage transformers are used to measure the line voltages on a three-phase system the secondaries are star connected giving line voltages of $63.5 \times \sqrt{3} = 110$ V at the meters.

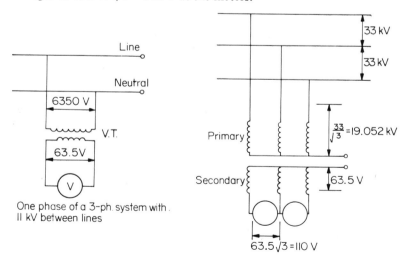

Figure 6.19   Figure 6.20

MEASUREMENT OF POWER AND POWER FACTOR

Current and voltage transformers are used to isolate wattmeters from the high voltage system in which the power is to be measured. The connections for a single-phase wattmeter are shown in *Figure 6.21*. The voltage and current coils are connected on one side to earth for safety reasons as already outlined.

Example (13).   A voltage transformer of ratio 100:1 and a current transformer of ratio 100:5 are used to measure the power and power factor in a single phase circuit using a wattmeter connected as shown in *Figure 6.21*. The potential difference across the wattmeter voltage coil is 63.5 V and the current in the current coil is 4.3 A. The wattmeter reading is 245 W.

Calculate for the primary circuit: (a) the current (b) the phase voltage (c) the power factor (d) the power.

(a) The C.T. ratio = 100:5

With 4.3 A in the secondary, the primary current = $4.3 \times \dfrac{100}{5}$

= 86 A.

(b) The voltage transformer ratio = 100:1
Primary phase voltage = $100 \times 63.5 = 6\,350$ V.

(c) The power factor in the secondary circuit is the same as that in the primary circuit assuming perfect transformers.

$$\text{Power factor} = \frac{\text{Power}}{\text{Volt-amperes}} = \frac{245}{63.5 \times 4.3} = 0.897.$$

(d) Power in the primary = $VI \cos \phi$ watts
$$= 6\,350 \times 86 \times 0.897$$
$$= 490\,000 \text{ W}.$$

This is the same as secondary power × V.T. ratio × C.T. ratio

$$= 245 \times \frac{100}{1} \times \frac{100}{5} = 490\,000 \text{ W}.$$

Example (14).   A load of 25 kVA with power factor 0.6 lagging is fed at 450 V from a single phase supply.

A voltage transformer with ratio 4:1 and a current transformer with ratio 100:5, together with a wattmeter, are used to measure the power in the circuit. Determine: (a) the potential difference across the wattmeter voltage coil (b) the current flowing in the wattmeter current coil (c) the indicated power on the wattmeter.

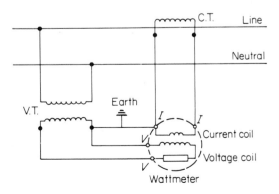

Figure 6.21

PROBLEMS FOR SECTION 6

(15) What is the purpose of the open circuit test on a transformer?

(16) What is the purpose of the short circuit test on a transformer?

(17) The core loss of a 240 V:110 V, 5 kVA single-phase transformer is 100 W when the 110 V winding is on open circuit and the 240 V winding is fed at 240 V. What will be the value of the core loss when the transformer is supplying its full rated load of 5 kVA?

(18) A transformer has a copper loss of 50 W when carrying one half full load. Calculate the value of the copper losses at (a) 70% of full load (b) full load.

(19) A transformer has a core loss of 100 W. What will be the value of the copper losses when the transformer is operating at maximum efficiency?

(20) Calculate the efficiency of a single phase transformer on full load at 0.8 lagging power factor given the following data.
Ratio, 500 V:100 V   Rating 20 kVA
Short circuit test results. 100 V winding short circuited. Power input to 500 V winding = 35 W. Current in 500 V winding = 20 A. Open circuit test results. 500 V winding open circuited. 100 V applied to the 100 V winding. Input 5 A at 0.2 power factor lagging.

(21) State three advantages of the auto-connection for transformers.

(22) State two disadvantages of the auto-connection for transformers.

(23) Estimate the ratio of copper in an auto-transformer to that in a double-wound transformer for ratios (a) 400 kV:132 kV (b) 400 kV:275 kV.

(24) Explain the possible effects of open circuiting the secondary of a current transformer when the primary is carrying current.

(25) Explain why the secondary winding of a current transformer is earthed.

(26) A 50 kVA, 11 kV:240 V single phase transformer draws 1.2 A at 0.3 power factor lagging from the 11 kV system with the 240 V winding open circuited. For the transformer on full load at 0.75 power factor lagging, calculate: (a) the secondary current (b) the primary balancing current (c) the primary power factor (d) the primary current.

7 D.C. machines

Aims: At the end of this section you should be able to:
Describe the construction of a d.c. machine.
Describe the principle and construction of armature windings.
Describe the function of a commutator and the process of commutation.
Derive the e.m.f. equation for a d.c. machine.
Understand that a reversal of armature current is necessary for a reversal of energy flow so that $E = V \pm I_a R_a$.
Discuss the connection between magnetic flux, armature current and torque.
Describe the necessity for and the operation of a d.c. motor starter.
Describe the losses which occur in, and draw a power flow diagram for, a d.c. machine.

PRINCIPLE OF ACTION

From Faraday's original work we know that $e = d\Phi/dt$ volts for a single turn where e = electromotive force in volts, $d\Phi$ = flux linked in webers and dt = time interval in seconds.

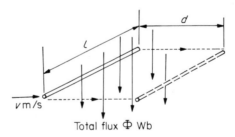

Total flux Φ Wb

Figure 7.1

Figure 7.1 shows a conductor of length l metres moving through an area of uniform magnetic field. The total flux is ΦWb and the conductor moves a distance of d metres at a uniform velocity of v m/s.

$$\text{Time taken} = \frac{\text{distance}}{\text{velocity}} = \frac{d}{v} \text{ s}$$

Total flux linked = flux density × area of the field.

$$= B \times l \times d \text{ webers}$$

$$e = \frac{\text{flux linked}}{\text{time}} = \frac{B \times l \times d}{d/v} = Blv \text{ volts.}$$

If the ends of the conductor are joined to an external circuit a current i amperes will flow.

Power = ei watts   Work done = power × time = ei × time

Therefore, work done = $\left(\dfrac{\text{flux linked}}{\text{time}}\right) \times i \times$ time

= *Bldi* joules.

But work done = force × distance operated through

Therefore, *Bldi* = force × *d*

Force = *Bli* newtons.

Thus to generate an e.m.f. the requirement is for magnetic flux, a conductor and relative motion between them. To produce a force on a conductor the requirement is for magnetic flux and a current flowing in that conductor.

In generators the flux is usually provided by a system of electromagnets and the motion of the conductors produced using an engine.

In motors the current is provided from an external source. This current both excites the electromagnets and feeds the conductor system.

In the d.c. motor the force created by the current flowing in the conductor system causes them to accelerate if they are free to move. The motion of the conductors through the field causes an e.m.f. to be generated. By Lenz's law, the generated e.m.f. opposes the supply voltage. For this reason it is called the back e.m.f. At a certain speed dependent on the mechanical forces involved, the strength of the magnetic field and the applied voltage, equilibrium will be reached between the applied voltage and the back e.m.f.

The applied voltage is given the symbol *V*
The back e.m.f. is given the symbol *E*.

In order that current shall flow into the motor to produce the necessary force and torque to sustain rotation, *E* must always be less than *V*.

$V - E = IR$ where the conductor current = *I* amperes and the conductor resistance = *R* ohms.

If the mechanical load on the motor is increased it will slow down so reducing the velocity of the conductors through the field. This reduces the back e.m.f. so causing the current to increase. The increase in current provides the extra force necessary to keep the motor turning against the increased load.

POWER, FORCE AND TORQUE

Force = mass x 9.81

Mass (kg)

Figure 7.2

Figure 7.2 shows a pulley connected to a motor. It is arranged to wind up a rope, the downward force in which is *F* newtons. The force is provided by a suspended mass. In one revolution of the pulley the mass is raised through $2\pi r$ metres.

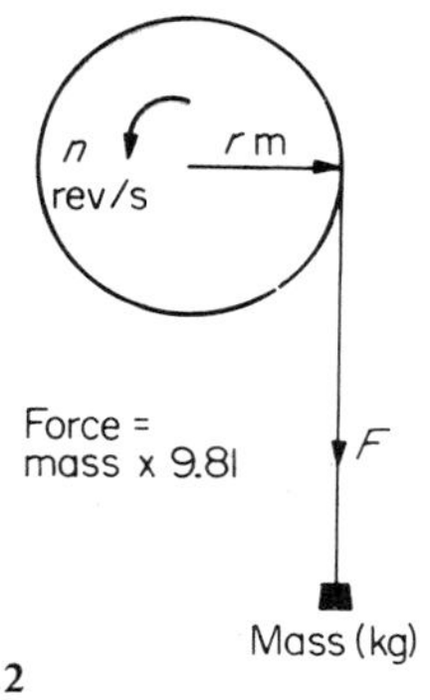

Work done = Force × Distance = $F \times 2\pi r$ newton metres (joules)

The pulley rotates at *n* rev/s

Therefore, work done per second = $F \times 2\pi r \times n$ J/s

Now 1 J/s = 1 watt and $F \times r =$ torque *T*

Hence power = $2\pi nT$ W

FLEMING'S RULES

For motors   The relationship between the direction of the force, current and magnetic field has been determined by experiment and a convenient way of remembering this is by the use of Fleming's left hand rule.

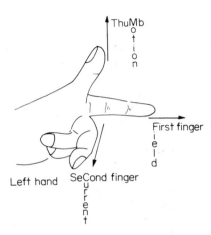

Figure 7.3 Fleming's left hand rule

In *Figure 7.3* the le*F*t hand relates current and magnetic field to *F*orce and force is produced by an electric motor to drive its load. The thu*M*b indicates the direction of the force and hence *M*otion The *F*irst finger represents the direction of the *F*ield The se*C*ond finger represents the direction of the *C*urrent

For generators   For *G*enerators the ri*G*ht hand is used. The thumb and two fingers represent the same quantities as for motors. In a generator the conductor system is driven by an external force and a current flows such as to produce a force which opposes that motion. This is clear from *Figures 7.4* and *7.5*. For the generator, the motion and field are unchanged but the current direction has reversed.

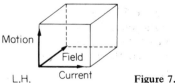 **Figure 7.4**   **Figure 7.5**

THE ACTION OF THE COMMUTATOR   Consider now the very simple generators shown in *Figure 7.6*. In both cases the magnetic fields are stationary and the coil rotates. *Figure 7.6a* is an alternating current generator and the output is taken from the coil using conducting sliprings on which blocks of carbon called brushes rub. Early machines used wire gauze made up like a sweeping brush and the name is still used. With the coil in the position shown the generated e.m.f. is zero since the coil sides are not cutting the lines of

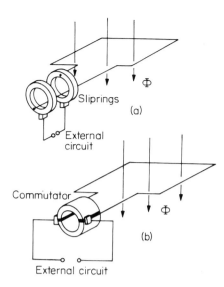

Figure 7.6

flux. As the coil revolves the voltage increases to a maximum after a rotation of 90° when the coil sides are moving directly across the flux. After a further 90° rotation the voltage will again be zero. As the rotation continues a voltage of reverse polarity will be produced, rising to a maximum and falling to zero as the revolution is completed since the direction of motion of the coil sides is reversed during this period. A coil side which moved from right to left during the first half revolution moves from left to right during the second half revolution.

Now consider the same coil but this time connected to a two segment commutator as shown in *Figure 7.6b*. This simple commutator is a split copper cylinder, each half fully insulated from its neighbour. As before, carbon brushes are used to connect the external circuit.

The operation of the commutator can be understood by looking firstly at *Figure 7.7*. The brushes are shown inside the commutator for clarity. Assume that there is an external load connected so that current flows in the coil. The left hand coil side connected to commutator segment 1 (hatched) has the current direction shown. This is deduced using Fleming's right hand rule. Current is flowing in the coil towards the left hand brush which delivers current to the external circuit and therefore has positive polarity. Current returns from the circuit to the right hand brush and flows in the coil away from the commutator segment.

In *Figure 7.8* we see the coil turned through 90°. The coil sides are outside the field so that the generated e.m.f. is zero. The brushes are shorting out the coil since they are touching both halves of the commutator at the same time. Later, when commutators with many more segments are considered it must be remembered that the change over of connection from one segment to the next must occur when the associated coil sides are out of the magnetic field.

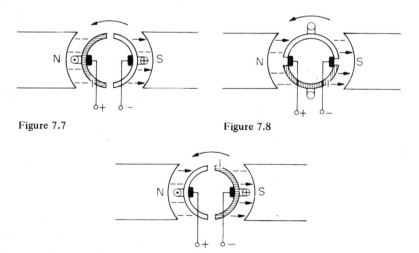

Figure 7.7   Figure 7.8

Figure 7.9

In *Figure 7.9* the coil has turned through a further 90° and commutator segment 1 is now on the right. The current direction in the coil side connected to it has reversed but since a commutator is being used the left hand brush is now connected to the other side of the coil and so still has the same polarity.

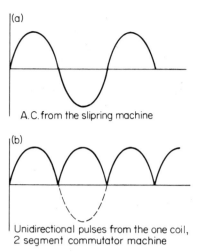

(a)

A.C. from the slipring machine

(b)

Unidirectional pulses from the one coil,
2 segment commutator machine

Figure 7.10

The e.m.f. generated in the coil is alternating but by using a mechanical reversing switch or commutator, the current flowing in the external circuit is a series of unidirectional pulses.

THE RING WOUND ARMATURE

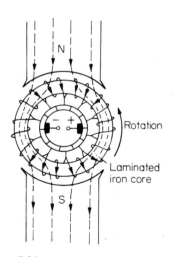

Rotation

Laminated iron core

Figure 7.11

In *Figure 7.11* the laminated iron core is wound with ten coils each having two turns. The commutator has ten segments, one for each coil. The magnetic flux produced by the poles crosses the air gap into the core. Only the outsides of the turns directly under the poles cut the flux and the directions of the induced e.m.f.s shown are again deduced using Fleming's right hand rule. In both the top and bottom halves of the winding this direction is towards the right hand brush which is therefore positive.

With the armature in the position shown each brush is shorting out one coil which has no e.m.f. induced in it since it is not cutting the flux. The flux at this instant is passing through the coil which is the required condition for commutation as already described (see *Figure 7.8*). Each coil passes through this position in turn as the armature rotates and during this instant carries no current.

There are six conductors cutting the magnetic flux from each pole at any instant in the arrangement shown in *Figure 7.11*. Representing the e.m.f. induced in each conductor by a cell, each half of the winding has six cells in series. Both halves of the winding are in parallel.

When a load is connected each half of the winding carries one half of the load current. As with the single coil, the direction of the e.m.f. induced in individual coils is alternating being in one direction as the coil passes under the north pole and in the opposite direction as it passes under the south pole. In this machine the air gap between the pole faces and the iron core is of uniform length so that the flux is uniform in the gap.

As a coil passes under a pole from point A to point B in *Figure 7.13*, the induced voltage is nearly constant. While moving from point B to point C the voltage falls to zero since no flux is being cut. The waveform of the voltage in a single coil is shown in *Figure 7.14*.

This should be compared with that shown in *Figure 7.10* which was for a different coil and field configuration.

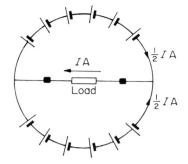

Figure 7.12

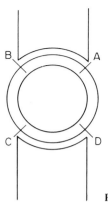

Figure 7.13

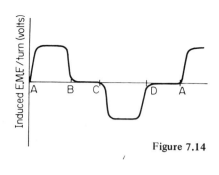

Figure 7.14

There are two disadvantages to the ring winding. These are:
1. The ring winding is difficult to wind since each turn must be taken round the core by hand.
2. Only a small part of the winding is active at any one time.
Other windings have been developed to overcome these problems.

THE LAP WINDING

Each coil in the lap winding overlaps its neighbours.

Figure 7.15 shows a lap coil comprising two turns. It is coil number 1 and its two sides are labelled 1 and 1′ respectively. Its two ends are connected to adjacent segments on a commutator and coil number 1 starts on segment number 1. The other end of the coil is connected to segment number 2.

Coil number 2 is added and this starts on segment number 2 of the commutator. Only one turn per coil is shown here making the diagram easier to follow. There may in fact be many turns in each coil.

The coils are situated in slots in a laminated iron armature and *Figure 7.17a* shows a part wound armature. *Figure 7.17b* shows how the windings overlap. The complete winding is built up in this manner until all the slots in the armature contain two coil sides, one on top which starts at the commutator segment which carries its number, 1, 2, 3 etc. and the other at the bottom which is the return coil side numbered 1′, 2′, 3′ etc. The fully wound armature rotates in a magnetic field which is generally produced electrically. A typical four-pole arrangement is shown in *Figure 7.18*.

Consider the typical section of a lap winding under a north and south pole as shown in *Figure 7.19*. The view is from the centre of the armature looking outward through the winding and seeing the pole faces outside the winding. The lines of force from the north pole are straight up out of the paper and those from the south pole down into the paper. Using Fleming's right hand rule the directions of the induced e.m.f.s are as shown. Between the positive and negative brushes there are six conductors connected in series so that the e.m.f. between the brushes is six times that induced in a single conductor. The complete lap winding has one brush for each pole and the brushes under like poles are connected together. For a four-pole machine there are four brushes and the arrangement is shown in *Figure 7.20*.

The winding shown in *Figure 7.19* is repeated four times between brushes A and B, B and C, C and D and D and A. The full winding for this simple machine therefore comprises 24 conductors (12 turns),

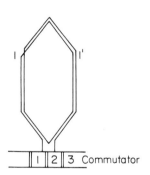

Figure 7.15

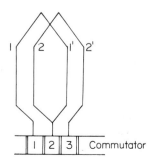

Figure 7.16

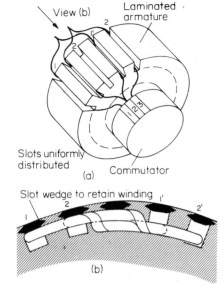

View (b)

Laminated armature

Slots uniformly distributed

(a)

Commutator

Slot wedge to retain winding

(b)

Figure 7.17

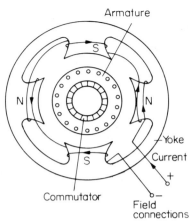

Armature

Yoke

Current

Commutator

Field connections

Figure 7.18

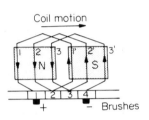

Coil motion

Brushes

Figure 7.19

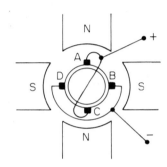

Figure 7.20

conductor 12′ returning to segment 1 under conductor 3 to complete the winding.

There are therefore four parallel paths through the armature:

From B to A, B to C, D to A and D to C.
Each path comprises six conductors in series.

Example (1).   A 6-pole d.c. lap wound generator has 36 slots on the armature. Each slot contains 8 conductors. When the speed of the armature is 500 rev/min, the induced e.m.f. in each conductor is 5 V.

Calculate: (a) the total number of conductors on the armature
           (b) the total number of turns on the armature
           (c) the number of conductors in series in each of the parallel groups.
           (d) the total generated e.m.f.
           (e) the rating of the generator if each conductor can carry 10 A before overheating.

(a) Total number of conductors = number of slots × number of conductors in each slot.

$$= 36 \times 8$$
$$= 288.$$

(b) Two conductors form one turn. Number of turns = $\frac{288}{2}$ = 144.

(c) Since there are six poles, there will be six brushes and six parallel paths through the armature.

Number of conductors in series in each group = $\frac{288}{6}$ = 48.

(d) Generated e.m.f. = e.m.f. per conductor × number of conductors connected in series.

$$= 5 \times 48$$
$$= 240 \text{ V}.$$

(e) Each conductor can carry 10 A and there are six parallel paths.

Total current from the armature = 6 × 10
= 60 A.

Rating = 240 × 60
= 14 440 W.

THE WAVE WINDING

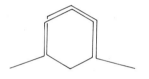

Figure 7.21

Figure 7.21 shows a coil for a wave winding. The ends of the coil are not connected to adjacent segments on the commutator but to segments some distance apart. Again using a four-pole machine as an example, the winding is shown in part in *Figure 7.22*.

Notice that commutator segments 2 and 3 are repeated at each end so that the commutator may be visualised in its circular form. There is an additional commutator segment, number 13, which causes the winding to progress in 'waves'.

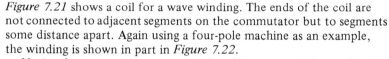
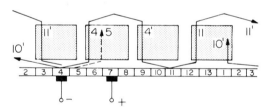

Figure 7.22

Starting with the negative brush presently on segment 4, follow conductors 4 and 4$'$ to segment 11, and then conductors 11 and 11$'$ to segment 5. Four conductors are involved in progressing one segment along the commutator. The process may be continued following four more conductors (not shown) so returning to segment 6. A further four conductors, making twelve in all allows us to progress to segment 7 upon which the positive brush rests.

With this winding two parallel paths through the armature exist. One of them is as described and the other could be traced from segment 4 once again starting with conductor 10$'$ which passes under the south pole on the right of the diagram, returning again to segment 7 eventually.

A wave wound machine has only two brushes and two parallel paths through the armature irrespective of the number of poles.

Example (2).   A four-pole wave wound generators has the following details:
Number of slots on the armature 64
Conductors per slot 15
Induced e.m.f. per conductor 1.8 V
Maximum current per conductor 15 A
Determine: (a) the output voltage (b) the maximum current which the machine may safely deliver.
(a) Total number of conductors = 15 × 64 = 960
There are two parallel paths, hence number of conductors in series per path = $\dfrac{960}{2}$ = 480.

Total induced e.m.f. = 1.8 × 480 = 864 V.
(b) Maximum current = 2 × 15 = 30 A.

Example (3).   A d.c. generator has 4 poles and 72 conductors on the armature. The e.m.f. generated per armature conductor at a particular speed is 10 V. The current carrying capacity of each conductor is 20 A. Calculate the output voltage and maximum power rating for (a) lap connection (b) wave connection.

THE E.M.F. AND SPEED EQUATIONS

Figure 7.23 shows part of a d.c. machine. Consider a single conductor on the armature as it moves from position A to position B.
Let Φ = total flux per pole in webers
  n = speed of the armature in revolutions per second.
  P = number of *pairs* of poles on the yoke. (Pairs are used since it is not possible to have a single pole.)
  Z = total number of conductors on the armature
  c = number of parallel paths through the armature.
There are $2P$ poles on the machine and at n rev/s the particular conductor shown will therefore pass $2Pn$ poles every second.

The time taken to pass one pole = $\dfrac{1}{2Pn}$ s

This is the time taken for the conductor to move from A to B in *Figure 7.23*. During this time the flux Φ from one pole is cut.

For a single conductor $e = \dfrac{\text{flux cut}}{\text{time (s)}}$ volts.

Hence the average voltage induced in the conductor = $\Phi / \dfrac{1}{2Pn}$

$$= 2Pn\Phi \text{ volts}$$

On the armature there are z/c conductors in series (see Examples 1 and 2).

Hence, the total e.m.f. $E = 2Pn\Phi\dfrac{z}{c}$ V

Transposing: $n = \dfrac{E}{2P\Phi\dfrac{Z}{c}} = \dfrac{Ec}{2P\Phi Z}$ rev/s

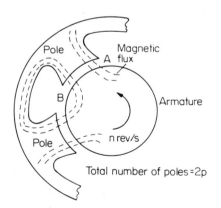

Pole
Magnetic
A flux
B
Armature
Pole
n rev/s
Total number of poles =2p

Figure 7.23

Since 1 revolution = 2π radians

$$\omega = \frac{2\pi Ec}{2P\Phi Z} = \frac{\pi Ec}{P\Phi Z} \text{ rad/s}$$

Example (4). The wave wound armature of a six-pole d.c. generator has 30 slots and in each slot there are 8 conductors. The flux per pole is 0.0174 Wb. Calculate the value of the e.m.f. generated when the speed of the armature is 1200 rev/min.

$$1200 \text{ rev/min} = \frac{1200}{60} = 20 \text{ rev/s}$$

For a wave wound armature $c = 2$
6 poles = 3 pole pairs. Hence $P = 3$.

$$E = 2 \times 3 \times 20 \times 0.0174 \times \frac{30 \times 8}{2} = 250.6 \text{ V}$$

Example (5). A lap wound d.c. generator is to have an output voltage of 500 V at 26 rev/s. The armature has 28 slots each containing 12 conductors. Calculate the required value of flux per pole.

For a lap winding there are the same number of parallel paths through the armature as there are poles.

Therefore $c = 2P$ and $E = \Phi nZ$ volts.
$$500 = \Phi \times 26 \times 28 \times 12$$
$$\Phi = \frac{500}{26 \times 28 \times 12}$$
$$= 0.057 \text{ Wb.}$$

D.C. MACHINE CONNECTIONS

Shunt *Figure 7.24* shows the shunt connection. The field winding comprises many turns of fairly fine wire and it is connected in series with a control rheostat directly across the supply. It is therefore in parallel with, or shunting, the armature.

    The field current is independent of the armature current unless this is large enough to cause a voltage drop in the supply lines in which case both the armature voltage and the field voltage will be affected.

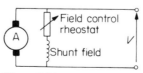

Figure 7.24

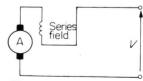

Figure 7.25

Series *Figure 7.25* shows the series connection. The field winding comprises a few turns of very heavy gauge wire or copper strip. It is connected in series with the armature and so carries the same current. For this reason it must have a very low resistance or the power loss will be excessive.

Compound *Figure 7.26a* and *b* shows machines with some shunt field and some series field. Both shunt and series coils are wound on the same pole pieces and the series field may either assist or oppose the shunt field. In the former case the machine is said to be 'Cumulatively wound' and in the latter case 'Differentially wound'. *Figure 7.26a* shows what is known as the short shunt connection whilst *Figure 7.26b* shows a long shunt.

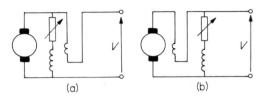

(a)   (b)

Figure 7.26

THE D.C. MOTOR

Speed/armature current characteristics When a voltage V is applied to the armature of a motor a current I_a flows. A force is produced which causes the armature to accelerate and a back e.m.f. E volts is generated. Clearly V must always be larger than E since if it were not so then current would not flow into the armature and no work could be done. The armature resistance $= R_a \Omega$.

Equilibrium is reached when $E = V - I_a R_a$ (7.1)

Transposing: $I_a R_a = V - E$ Therefore $I_a = \dfrac{V - E}{R_a}$

Now $n = \dfrac{Ec}{2P\Phi Z}$ rev/s.

For a particular machine P, Z and c are constants so let $\dfrac{2PZ}{c} = k$, a constant.

Now $n = \dfrac{E}{k\Phi}$ (7.2)

From equation 7.1 we see that any increase in I_a will cause the voltage drop $I_a R_a$ to increase and the voltage E to decrease.

In a shunt wound motor, if the field rheostat is not adjusted, the flux Φ will remain constant and from equation 7.2 since both k and Φ are constant, E and n are directly proportional.

In a series motor the field winding carries the armature current. The resistance of the armature circuit must include that of the series field since it will cause a voltage drop in addition to that of the armature itself.

Hence $E = V - I_a (R_a + R_{sf})$ where R_{sf} = resistance of the series field.

Ignoring the effect of magnetic saturation, the flux set up by the field winding is proportional to the armature current.

Again $n = \dfrac{E}{k\Phi}$ (7.2)

Now since increasing the armature current increases the flux, for a constant value of E the effect of increasing the load (and I_a) is to reduce the speed. Since in fact E falls as the armature current is increased (equation 7.1) there is a further reduction in speed due to this.

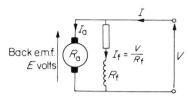

R_a = resistance of the armature
R_f = resistance of the shunt field
$I = I_a + I_f$

Figure 7.27

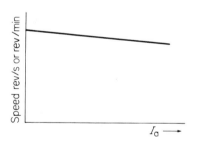

Figure 7.28

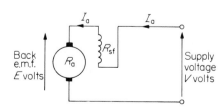

Figure 7.29

Example (6). A d.c. shunt motor has an armature circuit resistance of 0.5 Ω and a shunt field resistance of 240 Ω. It is connected to a 240 V supply. On no-load the input current is 2 A and the speed 1500 rev/min.

When fully loaded the input current is 8 A. Calculate the value of back e.m.f. generated and the speed of the motor at full load.

1500 rev/min = 25 rev/s

$I = I_a + I_f$ (from *Figure 7.27*) Therefore $I_a = 2 - 1 = 1$ A on no load.

$E_1 = V - I_a R_a$ (Using suffix 1 to indicate original no-load conditions)

$$= 240 - 1 \times 0.5$$

$$= 239.5 \text{ V}$$

$n_1 = \dfrac{E_1}{k\Phi}$ Transposing: $k\Phi = \dfrac{239.5}{25} = 9.58$.

When the total current = 8 A, $I_a = 7$ A (The field current remains constant)

$E_2 = 240 - 7 \times 0.5$ (Suffix 2 indicates the new fully loaded conditions)

$$= 236.5 \text{ V}$$

$n_2 = \dfrac{E_2}{k\Phi} = \dfrac{236.5}{9.58} = 24.68 \text{ rev/s}$

Figure 7.28 shows a typical speed/armature current characteristic for a shunt motor.

Example (7). A series motor has an armature resistance of 0.15 Ω and a series field resistance of 0.25 Ω. It is connected to a 250 V supply and at a particular load runs at 30 rev/s when drawing 10 A from the supply. Calculate the speed of the motor when the load is changed such that the armature current is increased to 20 A.

$E_1 = V - I_a (R_a + R_{sf})$

$$= 250 - 10(0.15 + 0.25) = 246 \text{ V}$$

As in Example 6, $k\Phi_1 = \dfrac{E_1}{n_1} = \dfrac{246}{30} = 8.2$

When $I_a = 20$ A, $E_2 = 250 - 20(0.15 + 0.25) = 242$ V

Now since the armature current has doubled, the flux has doubled.

$$n_2 = \frac{E_2}{k\Phi_2} \text{ where } k\Phi_2 = 2k\Phi_1 = 16.4$$

$$n_2 = \frac{242}{16.4} = 14.93 \text{ rev/s}$$

Thus as armature current increases the speed falls. *Figure 7.30* shows a typical speed armature current characteristic for a series motor. Notice that for low armature currents the speed is high since the flux is low and for this reason the series motor must always be connected to a load which will limit its top speed. On no load the speed would be sufficiently high to create disruptive centrifugal forces and the commutator segments and windings would be thrown outwards.

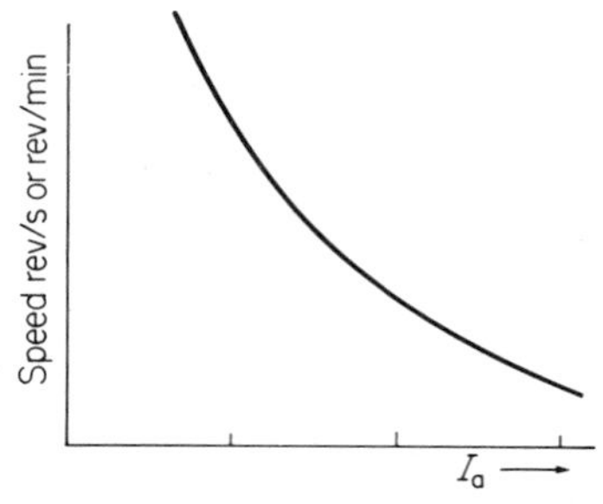

Example (8).   A lap wound armature for a four-pole d.c. machine has 56 slots each containing 10 conductors. For a flux per pole of 36 mWb, calculate: (a) the generated voltage E at a speed of 25 rev/s, (b) the speed at which the machine will run as a motor when drawing an armature current of 25 A from a 600 V supply given that the armature resistance is 0.5 Ω.

Example (9).   Repeat question (3) assuming that the armature is wave wound, all other conditions remaining unchanged.

Figure 7.30

A level characteristic may be produced by compounding. With the shunt motor the speed falls as the load is increased. By adding a small differentially wound series winding to the main poles the total flux per pole is reduced as the load current increases. Since speed is inversely proportional to flux, this causes an increase in speed compensating for the reduction due to the resistance of the armature.

Cumulative windings may be used to give a motor a characteristic between that of the series and shunt motors. The speed will fall as the load is increased but the motor will have a safe maximum speed on no load due to the constant shunt field.

The torque equation   From previous work we know that for a shunt motor

$$E = V - I_a R_a$$

Multiply throughout by I_a

$$EI_a = VI_a - I_a^2 R_a \quad \text{(all watts)}$$

VI_a = total power input to the armature (supply voltage × current)

$I_a^2 R_a$ = power loss in the armature as heat.

Hence EI_a must be the power available at the armature shaft to produce torque. (Note that by using $(R_a + R_{sf})$ throughout, the proof is valid for the series motor)

Now $E = \dfrac{2P\Phi Zn}{c}$ volts and power = $2\pi nT$ watts

Therefore $EI_a = \dfrac{2P\Phi Zn}{c} I_a = 2\pi n T$

Transposing $\quad T = \dfrac{2P\Phi Zn}{c2\pi n} I_a = \dfrac{P\Phi Z}{c\pi} I_a$ newton metres.

As before let $\dfrac{2PZ}{c} = k$

When $T = \dfrac{k}{2\pi} \Phi I_a$ Nm.   Where T = gross torque.

The gross torque developed is proportional to the flux and to the armature current. Not all this torque is available to drive an external load however since there will be friction in the motor bearings and between the armature surface and the air in the casing. The armature may be driving a fan which forces air over the windings to keep them cool and a torque is required to drive this. Finally, since the iron core of the armature is being driven through a magnetic field there will be hysteresis and eddy current losses.

Gross torque – all loss torques = net torque available to drive an external load.

Torque/armature current characteristics

In the shunt motor, for a constant rheostat setting, the flux Φ is constant.

Now $T = \dfrac{k}{2\pi} \Phi I_a$ Nm

Therefore I_a is the only variable and the gross torque is directly proportional to the armature current. This is shown in *Figure 7.31*.

In the series motor the field is produced by the armature current. $\Phi = k_1 I_a$ up to saturation where k_1 is the constant of proportionality between flux and current in webers per ampere. Substituting in the torque equation:

$$T = \dfrac{k}{2\pi} k_1 I_a^2$$

which is the equation of a parabola $[y = mx^2]$. This is shown in *Figure 7.32*.

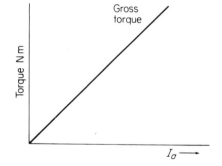

Figure 7.31 Shunt motor

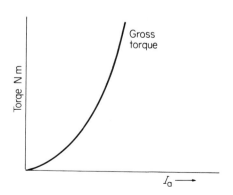

Figure 7.32 Series motor

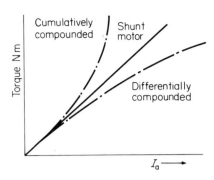

Figure 7.33

Differential compounding will cause the torque produced to be less than that of the shunt motor for a given value of armature current whilst cumulative compounding will increase the available torque.

Motor applications   Shunt motors are used where virtually constant speed is required on drives such as machine tools, fans and conveyor systems.

Series motors produce very high torques at low speeds and so they are suitable for starting very heavy loads. The main uses are in traction being used extensively in trains, electric buses, trams and milk delivery vehicles.

Compound wound motors are used where constant speed is required (differential winding) for coal feeders on boilers, and oil pumps, or where a high torque on starting and safe top speed are required (cumulative winding) for conveyor systems, hoists and cranes.

Speed control   A measure of speed control may be achieved by changing the field current of the motor. Since the induced voltage E is proportional to flux and speed any change in magnetic flux will result in a change in speed. The lower is the flux the higher is the speed. However torque is proportional to flux and armature current so that as the flux is reduced the armature current must increase if the load torque remains constant as the speed changes. This is often the limiting factor since large armature currents will cause overheating.

Usual values of speed obtainable by field weakening on a standard motor will be from normal up to something less than twice normal although a larger speed range is achievable using specially designed motors.

Since steel saturates at less than 2 tesla, increasing the field current and hence field strength as a method of speed reduction has very limited applications. This must be done by reducing the voltage on the armature. If the field and torque remain constant, the armature current and volt drop will remain constant.

Since $E = V - I_aR_a$, as the applied voltage V is reduced so the speed will fall in proportion in order to satisfy the e.m.f. equation.

Figure 7.34 shows characteristics of a shunt motor employing varying field currents.

In shunt motors a reduction in field current is achieved by using the field control rheostat (*Figure 7.24*).

In series motors a reduction in flux may be achieved by either by-passing part of the armature current round the field winding or by

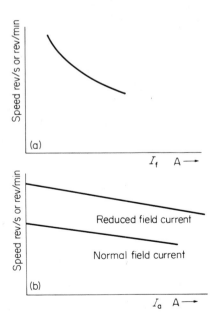

Figure 7.34

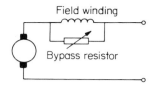

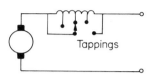

Figure 7.35

having a tapped field winding which allows a variable number of turns to be used. These two methods are shown in *Figure 7.35*.

Reduction of armature voltage can be obtained by using an additional resistor in the armature circuit but this is extremely wasteful of power since the whole of the armature current flowing in this resistor produces a considerable amount of heat. If the speed is reduced to one half normal by this method then one half of the input power is wasted.

Where large motors are involved controlled rectifiers or a d.c. generator may be used to provide the varying voltage. Silicon controlled rectifiers are often used when the mean output voltage can be controlled from zero to its full rated value at very high efficiency. A full description of the device is given in Chapter 13.

Where a d.c. generator is used, this is driven at constant speed, often by an a.c. motor. The output voltage from the generator may be varied from zero to full rated value by increasing its field current. By reversing the field connections the polarity of the output voltage is reversed. A motor supplied from this generator can be made to run at any speed from a crawl to its full rated value in either direction. Such a system is named after Ward-Leonard, the developer.

The shunt motor starter

Since $E = \dfrac{2P\Phi nZ}{c}$ volts

It follows that when a motor is at rest, since n = zero, whatever value of flux, the back e.m.f. is zero.

But $E = V - I_a R_a$ Therefore $0 = V - I_a R_a$

Transposing $V = I_a R_a$ or $I_a = \dfrac{V}{R_a}$ amperes.

The resistance of the armature is kept as small as possible to keep the losses in the armature to a minimum. It follows that the armature current under these conditions will be extremely large. It could in fact be large enough to cause severe damage to the armature by heating the conductors and commutator. Also since there are mechanical forces between current carrying conductors, with large currents sufficient force may be developed to disrupt the winding.

To prevent these excessive currents flowing, the resistance of the armature circuit is increased during the starting period.

In *Figure 7.36* the resistance R ohms is included in the armature circuit so that at standstill

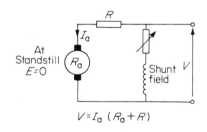

$V = I_a (R_a + R)$

Figure 7.36

$$0 = V - I_a(R_a + R) \text{ or transposing: } I_a = \frac{V}{(R_a + R)}$$

The added resistance is made large enough to limit the starting current to between 1.5 and 2 times the normal full load current of the motor. As the motor accelerates the back e.m.f. rises from zero so that

$$I_a = \frac{V - E}{(R_a + R)}.$$

The value of the added resistance can now be decreased somewhat to allow the armature current to rise to its original value once more. Further acceleration and an increase in the back e.m.f. will take place.

Reductions in additional resistance are made at intervals as the speed of the motor builds up. Finally when the armature resistance alone remains

$$I_a = \frac{V - E}{R_a}$$

During the starting process it is essential to have the maximum possible field strength in the motor since both torque and back e.m.f. are proportional to the field flux. The additional resistance must therefore only affect the armature circuit whilst the field winding is connected directly across the supply.

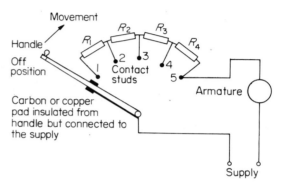

Figure 7.37

Figure 7.37 shows the armature circuit only of a d.c. shunt motor with a suitable starter. The handle is moved manually from the off position to make contact with the first resistance stud. This puts resistances R_1, R_2, R_3, and R_4 in series with the armature. As the motor speed rises from zero to a few revolutions per second the handle is moved so that the supply is fed to contact stud number 2. Resistances R_2, R_3 and R_4 are now in series with the armature. As the speed builds up the handle is moved successively across the studs ending up on number 5 when the armature is connected directly to the supply. As

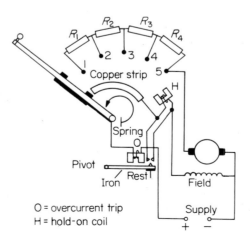

Figure 7.38

stated previously, the field current must be a maximum during this period and a method of achieving this is shown in *Figure 7.38*.

The starting handle being moved on to the first stud puts all the resistances in series with the armature. The current from the supply positive terminal flows through *O*, the overcurrent trip to the bottom of the starting handle along which there is a conducting strip to two contact pads. One of these feeds current to the series resistors whilst the other makes contact with the copper strip which is connected to the field winding. On the way to the field winding the current flows through *H*, the hold-on coil. The field winding is therefore subjected to the full supply voltage (less a very small voltage drop in coil *H*). As the starting handle progresses to the running position the field remains unchanged.

When the handle reaches stud 5 it comes into contact with the faces of the electromagnet *H*. Either the whole handle or just a small pad fixed to it is made of a magnetic material so that provided that a suitable current is flowing in the field winding, the handle will be held in the running position.

All the current being supplied to the motor flows through the small coil on the horseshoe shaped piece of iron of the overcurrent trip *O*. A beam beneath the coil is held on the rest stop by gravity or a small spring. Normally the magnetic effect of the current is too small to attract the beam upwards. If however the current becomes excessive the beam is pulled upwards and the two electrical contacts to the right are made which causes the field current to be bypassed round the hold-on coil. This becomes de-energised and the starting handle is released. A spring causes it to fly back to the start position.

If at any time the supply to the motor is lost, the coil *H* will be de-energised and the handle will return to the start position. This prevents the motor from starting up unexpectedly when the supply is restored with possibly disastrous results both to the motor and to the operator who may have a hand in the driven equipment.

THE D.C. GENERATOR

Shunt wound

If a d.c. machine connected to a supply and running as a motor, has the speed of its armature raised above that at which it operates as a motor by driving it with an external engine whilst keeping the field constant, the value of the generated e.m.f. will increase since this is proportional to speed.

As E becomes greater than the supply voltage a current will flow from the machine which has now become a generator, receiving power from the driving engine and delivering power to the electrical system. The difference between E and V is still $I_a R_a$.

$$E = V + I_a R_a.$$

For a motor $E = V - I_a R_a$ and the sign change for the generating condition indicates a change in the direction of the current. In the self-excited generator in *Figure 7.39* part of the armature output is used to excite its own field so that $I = I_a - I_f$. An increase in load current causes $I_a R_a$ to increase and the terminal voltage V will fall.

The field current $I_f = V/R_f$ so that a reduction in terminal voltage results in a reduction in the field current and hence the field flux.

Since $E = k\Phi n$ volts, the reduction in flux causes a reduction in

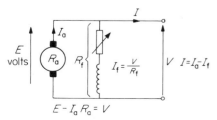

Figure 7.39

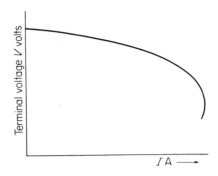

Figure 7.40

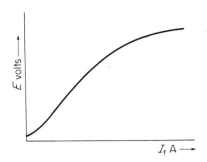

Figure 7.41

generated e.m.f. The terminal voltage of a self-excited generator there-fore falls quite rapidly as the load current increases due to the combined effects. A typical characteristic is shown in *Figure 7.40.*

When driven at constant speed with no load connected, the relation-ship between field current and generated e.m.f. may be determined using the circuit shown in *Figure 7.39* with the addition of a voltmeter to measure output voltage and an ammeter in the field circuit. Due to magnetic saturation the output voltage is not proportional to the field current. With zero field current there is a small output voltage which is produced by the residual magnetism in the pole pieces without which the generator cannot commence generation. In a new machine this is created using an external power source. *Figure 7.41* shows the open circuit characteristic.

Series wound

This is rarely if ever used since with no load on the machine there is no current in the field winding and the output voltage is near zero. When a current flows a flux is produced and the output voltage rises. The terminal voltage is therefore a function of the load current.

Compound wound

Cumulative compounding is used to overcome the drooping voltage characteristic shown in *Figure 7.40* for the shunt machine. As load current flows in the series winding a flux is produced which is added to that of the shunt field. This increase in total flux produces a correspond-ing increase in generated e.m.f. The output voltage can thus be held constant or made to increase slightly to compensate for the voltage drop in the connecting cables to the external load.

Differential compounding is used to cause the terminal voltage to collapse when load current flows. Load current increases the series field which opposes the shunt field so causing a reduction in flux and generat-ed e.m.f. This is used in electric arc welding sets where approximately 110 V is required to strike the arc but only 20 V to maintain the arc whilst welding.

ARMATURE REACTION

Further work on commutation

Figure 7.42a shows one coil of the lap winding originally shown in *Figure 7.19.* The negative brush is resting on segment 4 of the com-mutator. As viewed the current direction in the coil 3, 3′ is anticlockwise. In *Figure 7.42b* the conductors have moved a distance equivalent to one half of a commutator segment to the right. Both coil sides are now

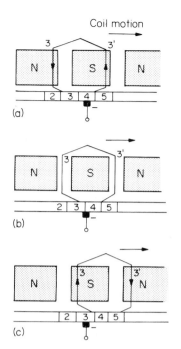

Figure 7.42

between poles and there is no induced e.m.f. in the coil. The brush now shorts out the coil since it spans segments 3 and 4.

In *Figure 7.42c* the coil has again moved to the right a further half segment of the commutator and both coil sides are under poles. The currents are now as shown, the direction being clockwise as viewed.

This means that in the time taken for the armature to move from position (a) to position (c) the current in the coil must reverse in direction. If the coil has a large inductance which is likely since it is embedded in a large mass of iron, and again if the current is large, there will be a considerable induced voltage in the coil during the period of reversal. This is known as reactance voltage.

The reactance voltage

$$e = L \frac{di}{dt} \text{ volts}$$

where L = inductance of the circuit in henrys and di/dt = rate of change of current in amperes per second.

Severe sparking will occur at the surface of the commutator if the current reversal, or commutation, is not complete by the time that the commutator segment moves from under the brush.

During commutation the sides of the coil must be outside the magnetic field so that the generated e.m.f. is zero and to achieve this the brushes are moved round the commutator until they are located on what is termed the neutral axis.

Armature reaction

In *Figure 7.43a* a simple armature is shown revolving between two magnetic poles. The current directions are as shown. The coil A,A′ which lies horizontally has no e.m.f. induced in it and the current in it is undergoing commutation. The coil B,B′ which is in a vertical plane at this instant is shown separately in *Figure 7.43b*. The current in conductor B flows away from the commutator end along the length of the armature, crosses over at the back and then flows back in conductor B′. This current produces a magnetic north pole on the left hand side of the armature. The other two current carrying coils will add to the effect. The strength of this cross-magnetic field will depend on the number of turns and the value of armature current.

In *Figure 7.44*, two ways of illustrating the effect of the armature or cross flux are shown. In *Figure 7.44a* the distorting effect of the current in the two sides of the coil B,B′ is shown. In *Figure 7.44b* the vector addition of the armature and main fluxes can be seen. The resultant flux has been displaced by $\theta°$ from the vertical. The effect of the armature flux on the main flux is called armature reaction. This term must not be confused with reactance voltage which is quite different.

In a multi-pole machine where the space between the poles is much smaller than in the two-pole machine this twisting of the field can bring into the field conductors which were out of it and in which commutation is taking place. Severe sparking results. The position for the brushes to be situated in order to obtain satisfactory commutation has now moved $\theta°$ in a clockwise direction from the former position. The new position is known as the magnetic neutral axis and its position is a function of the armature current since the amount of field distortion is dependent on this current. In some older machines brush moving gear

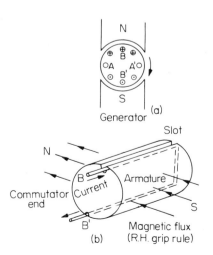

Figure 7.43

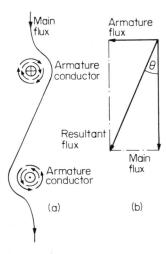

Figure 7.44

was provided so that an operator could adjust the position of the brushes as the load changed. This is not always convenient and other methods are generally employed to achieve satisfactory commutation under all load conditions.

Compensating windings and interpoles

One method of overcoming the effect of armature reaction is to fit compensating windings to the machine. These carry the armature current and so are series connected. They are arranged to create magnetic poles which oppose those set up by the armature currents so that at any instant the resultant cross field is zero. This type of winding is very expensive and is therefore only used on some special machines.

A more usual alternative method is the use of interpoles. These are very thin poles carrying armature current and these are situated

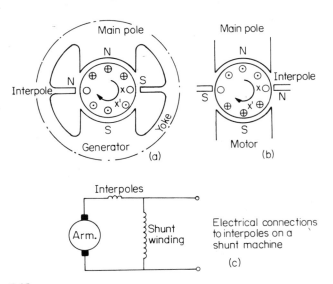

Figure 7.45

between the main poles. They are wound to give a polarity which anticipates the action of the next main pole. They assist the reversal of the current in the coil undergoing commutation. The interpole not only neutralises the effect of armature reaction local to the conductors undergoing commutation but overcomes the effects of reactance voltage.

When considering the action of the interpoles it must be remembered that they always act in the generating mode whether they are fitted to motors or generators. Fleming's right hand rule always applies to them. They assist in inducing a voltage of the correct polarity before the conductor actually moves under the next main pole. In *Figure 7.45a* the simple generator rotating in a clockwise direction has the armature current directions as shown. As the conductor on the right moves from a position X to X′ the current must be established out of the paper as viewed. In order to assist this to occur the interpole field must be as shown. *Figure 7.45b* shows the interpole polarities for a motor with the same rotation.

Changes in characteristics due to armature reaction

Armature reaction causes the magnetic field in a d.c. machine to be twisted round which concentrates the flux into one side of each pole piece so that this side may be driven into saturation as a result. Since the top limit to flux density in iron is in the region of 1.6 to 1.8 tesla this may result in a reduction in total flux.

Example (10). The normal flux density of the poles in a d.c. generator on no load is 1.1 T. Each pole has a cross-sectional area of 0.05 m^2. The effect of armature reaction at a particular load is to concentrate all the flux into one half of the pole face area. The saturation flux density is 1.8 T. Calculate the new value of flux per pole and the per unit reduction in flux at this load.

Total flux per pole = flux density × area of the pole face.
$$= 1.1 \times 0.05$$
$$= 0.055 \text{ Wb.}$$

Concentrating all this flux into one half of the pole face could be expected to give a flux density of

$$\frac{0.055}{0.05/2} = 2.2 \text{ T.}$$

This is above saturation. The flux density cannot rise above 1.8 T. At 1.8 T and an area of 0.025 m^2 the flux per pole = $1.8 \times 0.025 = 0.045$ Wb. This is a reduction of $0.055 - 0.045 = 0.01$ Wb.

$$= \frac{0.01}{0.055} = 0.19 \text{ p.u. reduction.}$$

In practice the reduction is seldom if ever this large, but the example serves to demonstrate the principle.

This reduction in flux will cause a further reduction in the output voltage of d.c. generators in addition to those already discussed.

In the case of motors there will be a speed increase as the flux is reduced and in addition, if the load torque remains constant, the armature current will increase. These effects can be explained by a consideration of the e.m.f. and torque equations.

LOSSES IN D.C. MACHINES

(See *Figures 7.27, 7.29* and *7.39* for connections and symbols.)

Field loss

In the shunt machine the field loss = VI_f or $I_f^2 R_f$ watts.
In the series machine, since the field winding carries the armature current, field loss = $I_a^2 R_{sf}$ watts.
The field loss is a function of the number of ampere turns necessary to magnetise the pole pieces and so is dependent on the grade of steel used.

Armature loss

(a) $I_a^2 R_a$ watts loss in the coils of the armature. For a particular load the armature current is constant so that this loss is directly proportional to the resistance of the winding.
(b) Iron losses. Since the armature is subjected to alternating magnetisation as it passes under north and south poles, there will be hysteresis and eddy current losses in the steel. The hysteresis loss is a function of: (1) the maximum flux density (2) the grade of steel used and (3) the speed of rotation of the armature. Eddy current losses are a function of: (1) the maximum flux density (2) the thickness of the laminations used (3) the resistivity of the steel and (4) the speed of rotation.

Commutator losses

(a) Resistance loss. There is resistance between the brush surface and the commutator and the loss depends on the grade and quality of the brushes used. The volt drop tends to be constant at about 1 V per brush so that this loss is approximately (2 volts × total armature current) watts, irrespective of the type of armature winding. (b) Friction loss. There is rubbing friction between the brushes and the commutator and the power loss depends on the coefficient of friction and the rubbing speed. The rubbing speed is a function of the diameter of the commutator and the speed of rotation.

Bearing friction

The armature is supported in bearings and these may be of the ball, roller or sleeve types according to motor application. The frictional loss is a function of the speed of rotation and the type of bearing.

Windage

There will be friction between the surface of the armature and the air in the casing. In addition power will usually be required to drive the cooling fan. The power loss involved depends on the type and size of fan and on the running speed.

THE POWER FLOW DIAGRAM

For a motor

Note: The term 'Armature circuit' includes any series field windings and interpoles present.

Starting from the top of *Figure 7.46* we have the total input to the motor. From this is subtracted the shunt field loss if applicable. VI_a watts is supplied to the armature circuit where the losses are I_a^2 (total resistance of the armature circuit) watts plus the brush resistance loss. The power to create the gross torque is EI_a watts. The gross torque drives the armature against brush and bearing friction, windage and the retarding torque due to hysteresis and eddy currents in the armature. After all the losses have been provided we have the useful output which drives the connected mechanical load.

For a generator   Starting at the bottom of *Figure 7.46* we have the total input which is being provided by an engine of some form. The input must provide all the mechanical losses leaving EI_a watts from which the armature circuit losses must be subtracted. This leaves VI_a watts as the output from the armature circuit. Finally in the shunt machine its own field loss must be deducted leaving the useful power output at the top.

In both the motor and the generator case the input is greater than the output and the ratio output/input is the efficiency of the machine.

Typically this will lie between 0.6 and 0.85 according to the rating of the machine.

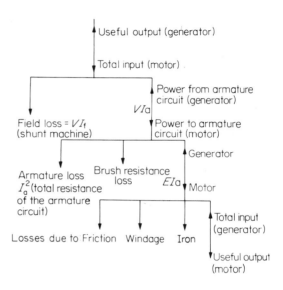

Figure 7.46

PROBLEMS FOR SECTION 7

(11) What is the function of a commutator in a d.c. machine?

(12) What is the basic difference between a lap and a wave winding on a d.c. armature?

(13) Which type of winding would you choose for: (a) high current, low voltage application (b) high voltage, low current application.

(14) Calculate the value of the generated e.m.f. delivered by the following six-pole d.c. generators.
(a) Lap wound: $\Phi = 20$ mWb/pole,
    number of slots on the armature = 86
    conductors per slot = 5, speed = 15 rev/s
(b) Wave wound: $\Phi = 25$ mWb, number of slots = 24,
    conductors per slot = 10, speed = 20 rev/s

(15) A d.c. machine has an armature resistance of 0.5 Ω. It is connected to 500 V mains. When drawing 20 A from the supply it runs at 25 rev/s. At what speed must it run as a generator in order to deliver 20 A to the electrical system? The flux remains constant.

(16) The armature of a six-pole, lap wound d.c. motor has 54 slots and 8 conductors in each slot. The total flux per pole is

0.05 Wb. The resistance of the armature is 0.3 Ω. When connected to a 240 V supply, at a particular load the armature current is 20 A. Calculate: (a) the speed under these conditions (b) the value of flux per pole required to increase the speed to 15 rev/s all other conditions remaining unchanged.

(17) Draw simple circuit diagrams illustrating (a) the series connection (b) the shunt connection and (c) a compound connection.

(18) Sketch speed/armature current and torque/armature current characteristics for motors (a) and (b) in Q 17 above.

(19) A d.c. shunt motor has an armature resistance of 0.3 Ω. It is to be started using a resistance starter. The supply voltage is 250 V. Determine the value of starting resistance for the armature circuit so that the current at standstill does not exceed 25 A.

If at a particular speed the back e.m.f. has risen to 125 V, to what value should the starting resistance be reduced to give the original value of current?

(20) What relationship must exist between armature coils and field poles in order to achieve good commutation? What design features are incorporated in many machines to assist commutation?

(21) A d.c. machine has the following parameters: $R_a = 0.5\ \Omega$ $R_f = 220\ \Omega$. Brush resistance loss = 100 W, Friction, windage and iron losses = 1500 W. Supply voltage = 440 V. Determine the output power and efficiency of the machine when operating as a motor and the armature current is 50 A.

(22) For the machine in Q.21 operating as a generator supplying 50 A to the electrical system, determine the total power input and efficiency.

8 Measurements and components

Aims: At the end of this section you should be able to:
Connect ammeters and voltmeters correctly in electric circuits.
Describe methods for the extension of meter ranges.
Describe the operation of an ohmmeter.
Explain the necessity for, and the limitations of rectifier instruments.
Compare moving-coil, moving-coil rectifier and moving-iron instruments.
Connect a wattmeter correctly in a circuit.

AMMETER AND VOLT-METER CONNECTIONS

The rate of flow of a current of electricity is measured in coulombs per second. A rate of flow of 1 coulomb per second is called the ampere. The rate of flow in amperes is measured using an ammeter which must be connected in series with any load so that the same current flows in both the ammeter and the load.

The potential difference between two points in a circuit is measured in volts.

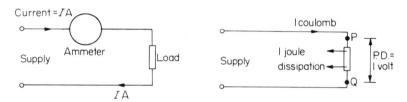

Figure 8.1   Figure 8.2

In *Figure 8.2*, a potential difference of 1 volt exists between points P and Q if 1 joule of energy is dissipated for each coulomb of charge which passes through the resistor. To measure the potential difference between two points in a circuit, the voltmeter must have connections to those two points. The voltmeter is connected in parallel with the component across which the potential difference is to be measured.

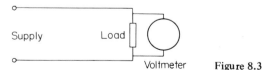

Figure 8.3

EXTENSION OF METER RANGE

The ammeter
The basic instrument for many electrical measurements is the permanent-magnet moving-coil ammeter. The movement is constructed using very fine wire so that it can pass only a very small current without overheating. By the addition of an external bypass resistor, or shunt, it may be used to measure larger currents.

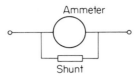

Figure 8.4

Example (1). A moving-coil ammeter has a coil resistance of 5 Ω and is fully deflected when a current of 2.5 mA flows through it.

Calculate the value of shunt required in order that the meter will be fully deflected when it is connected in a circuit which carries a current of 1 A.

When fully deflected the potential difference across the meter movement = *IR* volts

$$= 2.5 \times 10^{-3} \times 5$$
$$= 0.0125 \text{ V}.$$

This must never be exceeded or damage to the meter will occur. Total circuit current = 1 A. Therefore the shunt must carry $1 - 0.0025 = 0.9975$ A.

The potential difference across both meter and shunt are the same since they are connected in parallel.

$$\frac{\text{Potential difference}}{\text{Current in shunt}} = R_{\text{shunt}}$$

$$\frac{0.0125}{0.9975} = 0.0125 \ \Omega.$$

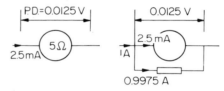

Figure 8.5

The voltmeter   Where the moving coil instrument is to be used as a voltmeter it is necessary to limit the current flowing through it to that value required to give full-scale deflection.

This is done by connecting a resistance, or multiplier, in series with it as shown in *Figure 8.6*.

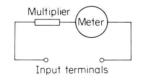

Figure 8.6

Example (2). For the same meter as in Example 1, determine the value of a multiplier to be fitted to enable it to be used as a voltmeter with a range of 0–50 V.

The current through the movement must not exceed 0.0025 A. With applied voltage 50 V, the total circuit resistance must be

$$\frac{50}{0.0025} = 20\,000 \ \Omega.$$

The resistance of the movement itself = 5 Ω.

The value of the multiplier is therefore $20\,000 - 5 = 19\,995 \ \Omega$.

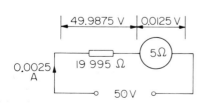

Figure 8.7

Meters also have their ranges extended by the use of instrument transformers and this method is described in Chapter 6.

THE OHMMETER

The ohmmeter comprises an ammeter with its face calibrated in ohms, a battery and calibration resistances. As an example let us assume a battery voltage of 4 V and the meter to give full-scale deflection for a current of 0.05 A.

To set up the ohmmeter, the external leads from the terminals P and Q in *Figure 8.8* are connected directly together. The resistance being measured is now zero. The ammeter is to indicate maximum current under these conditions. Since the battery e.m.f. is 4 V

$$\text{Total circuit resistance including leads} = \frac{4}{0.05} = 80 \ \Omega.$$

Resistor R_y is adjusted until the meter indicates 0.05 A. This corresponds to zero external resistance and is marked 0 on the resistance scale.

Disconnecting the leads results in zero current flowing and the ammeter scale zero is marked 'Infinity' (∞) Ω.

The scale is calibrated in ohms between zero and infinity. For example, when 0.04 A is flowing, the total resistance = 4/0.04 = 100 Ω. Since all internal resistances and leads accounts for 80 Ω, the external resistance must be 20 Ω. 0.04 A on the ammeter current scale corresponds to 20 Ω on the resistance scale.

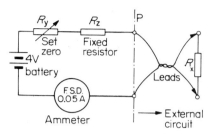

Figure 8.8

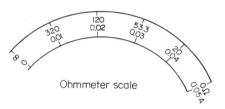

Ohmmeter scale

Figure 8.9

Figure 8.9 shows the complete scale. Notice that it is not linear and the indication may be difficult to interpret when measuring high resistance values.

Having set the zero ohms position correctly using the resistor R_y, the meter will read directly the resistance value of any external resistance connected to the leads. If the zero cannot be set using R_y, low battery voltage is indicated.

The meter ranges are altered in the multirange instruments by (a) changing the battery voltage and (b) fitting the meter with a shunt whilst at the same time changing the value of R_z.

RECTIFIER INSTRUMENTS

The deflection of the permanent-magnet moving-coil instrument is proportional to the average current flowing in the movement. In a d.c. circuit, connecting the meter terminals correctly causes the pointer to move up the scale while reversing the connections results in the pointer being forced back on the zero stop.

If the meter is connected into a circuit carrying alternating current the meter indicates zero since the average value over a complete cycle of a symmetrical alternating current is zero. During the positive half cycle there will be a torque on the meter movement in one direction whilst during the negative half cycle the direction of the torque will be reversed.

In order to employ a moving-coil meter in an a.c. circuit it is necessary to connect it in series with a rectifier.

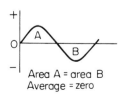

Figure 8.10

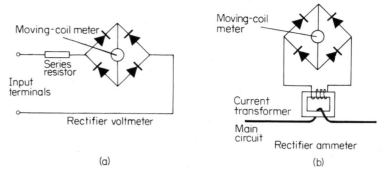

Figure 8.11

(a)

(b)

Figure 8.11a shows the use of a single-phase bridge rectifier in a voltmeter circuit. The series resistor is used in exactly the same manner as in the direct current circuit to limit the current. *Figure 8.11b* shows the connections for a rectifier ammeter. A current transformer is generally used to reduce the current to a suitably low value for the meter and rectifiers to handle being typically in the milliampere range. The alternative would be to use a meter shunt whilst the rectifiers pass the total circuit current.

Assuming a sinusoidal supply, *Figure 8.12* shows the ideal wave shape of the current in the meter for the connections in *Figure 8.11.*
For a sine wave, the average value of this current is $0.636\,I_{max}$.
The r.m.s. value of the current = $0.707\,I_{max}$.
The r.m.s. value is therefore 1.11 × the average value.

Figure 8.12

The deflection of the meter is due to the average torque which is a function of the average current. The meter must however indicate the effective or r.m.s. value so that it is necessary to re-scale the meter face, increasing all the values by a factor of 1.11. The scale markings are therefore only correct if the input to the meter is of sinusoidal form.

ERRORS

Due to waveform Where iron cored inductors, transformers or semiconducting devices are involved it may well be that waveforms are not sinusoidal and there will be errors in the meter indication when using a rectifier instrument. As an example, consider the effect of connecting a voltmeter to a supply of triangular wave shape with a maximum value of 100 V.

The average value of a triangular wave is one half of the maximum and the meter scale is increased by a factor of 1.11.

The meter indicates (0.5 × 100) × 1.11 = 55.5 V.
The r.m.s. value of the triangular wave = 57.73 V.

The moving coil meter therefore indicates a value which is 2.23 V low. This is an error of 2.23/57.73 = 0.0386 p.u. (3.86%).

In addition to non-sinusoidal wave shapes fed to the meter from external sources, the meter rectifier itself produces distortions of the waveshape.

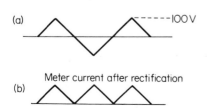

(a)

- - - - - 100 V

Meter current after rectification

(b)

Figure 8.13

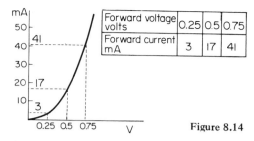

Figure 8.14

Figure 8.14 shows the characteristic of a rectifying device with a table of values taken from it. If a voltage of sinusoidal form and peak value 0.75 V is applied to the rectifier, the output terminals of which are connected to a low resistance meter, the current will have a peak value of 41 mA and the waveshape will be distinctly non-sinusoidal. The voltage and current waves are shown in *Figure 8.15*. These results show that even when the voltage applied to a rectifier meter is sinusoidal the current in the meter movement is not and waveform errors occur. For voltmeters measuring up to 10 V the error may be significant but in higher range voltmeters the series resistor is large and swamps the effect of the rectifier. This error is not relevant to ammeters, the current waveshapes being determined by the load circuit.

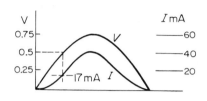

Figure 8.15

Due to frequency

The anode and cathode of any rectifying device make up a small capacitor since they are sections of conducting material separated by a insulator. At power frequencies up to possibly 5 kHz, the value of capacitive reactance is sufficiently high that little charging current flows from the supply.

Capacitive reactance $X_C = \dfrac{1}{2\pi fC}$ Ω, and charging current $I_c = \dfrac{V}{X_C}$ A.

At very high frequencies the falling capacitive reactance becomes significant since it allows charge to move back round the circuit during reverse half-cycles.

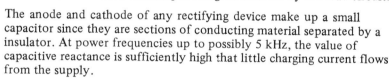

Figure 8.16

Considering the half-wave rectifier shown in *Figure 8.16*, during the positive half cycles of voltage the rectifier conducts and the potential difference across the rectifier is small.

$V_{max.} = I_{max.} \times R$ (Closely)

During the negative half-cycles, ideally the current is zero and the potential difference across the rectifier is $-V_{max}$. This voltage causes charge to move round the circuit in the reverse direction, charging the capacitance of the device. The current in the reverse direction will reduce the reading on the meter. The error involved depends on the capacitance of the rectifying devices and this differs with type. Copper oxide rectifiers have a relatively high capacitance and an error of 10% could be expected at 100 kHz. Point contact diodes have very low capacitance.

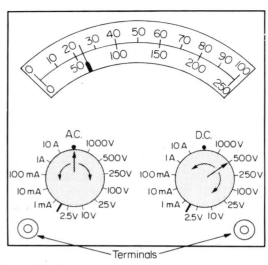

Figure 8.17

USE OF THE MULTIMETER

There are many different makes of multimeter so that it is possible only to give general guidance as to their use. Generally the meter has two selection knobs and the meter face has at least two different scales covering typical ranges from zero to 1000 V, 500 V, 250 V, 100 V, 25 V, 10 V and 2.5 V and from zero to 10 A, 1 A, 0.1 A and 0.001 A; all ranges for either a.c. or d.c.

For the meter shown in *Figure 8.17*, if an alternating voltage of unknown value is to be measured, the left hand knob is turned so that its arrow is in the a.c. position. The right hand knob is turned to the 1000 V range. With this range the scale to be observed is that from 0–100 and each of the scale values must be multiplied by 10. As the supply to be measured is connected to the meter terminals the pointer will move up the scale. If it moves to a position less than 50 on the scale (less than 500 V applied) the right hand knob should be turned one position clockwise so that the 500 V range is selected. This position is shown in the diagram. Either scale is now suitable. The top scale multiplied by 5 or the bottom scale multiplied by 2 for the actual value. The voltage selector knob is turned one position at a time until ideally the pointer is deflected to near full scale. Care must be taken not to select a range which would cause the pointer to deflect past the end of the scale since coming into contact with the end stop will bend the pointer and may damage the movement. Severe overloads can overheat the coil. Some meters incorporate a cut-out which disconnects the meter from the supply under these conditions but often this occurs too late to prevent damage. The reading is recorded taking care to allow for any scale factors involved.

Where alternating current is to be measured the left hand knob is left on a.c. and the right hand knob selected to the highest current range, in this case 10 A. The range is lowered one position at a time as before.

The same procedures apply to the measurement of direct voltage and current, the right hand knob being selected to d.c.

These are only typical values and positions of the knobs etc. and the particular instrument to be used together with any instruction material must be studied carefully before attempting to connect a meter which is strange to you, to the supply.

COMPARISON OF METER TYPES

To assist in the understanding of the accompanying table, let us first be clear on the following terms:

Deflecting or driving torque.   This is the torque which causes the meter movement to be deflected so driving the pointer up the scale.

Restraining torque.   This torque acts in opposition to the driving torque. Without it the smallest current in the meter coil would cause full scale deflection of the pointer and when the current ceased the pointer would not return to the zero position. The pointer comes to rest over the scale when driving torque = restraining torque.

Damping torque.   This is necessary to prevent undue oscillation of the pointer about the true reading. Damping torque is only present whilst the pointer is moving. Ideally the meter should be 'Dead-beat'. This means that when the supply is connected to the meter, the pointer moves steadily up the scale coming to rest at the correct reading, or possibly passes the correct point once and comes to rest on the way back.

Table 8.1 . Comparison between moving-coil and moving-iron meters

	Moving-coil meter	*Moving-iron meter*
Deflectional or driving torque	Current flowing in a coil of fine wire suspended between the poles of a permanent magnet	Usually repulsion between a fixed and moving iron. The current is carried by a stationary coil which surrounds the irons
Restraining torque	Hair springs	Hair springs
Damping torque	Electrodynamic. The coil is wound on a conducting former. As the coil moves the former is driven through the magnetic field so acting as a generator. Current flowing in the former produces a retarding torque which ceases when the coil comes to rest	Vane or fan moving through air in a cylinder or dash pot
Other comments	D.C. only, the movement carries at least part of the current being measured. Very sensitive, full-scale deflection for only a few microamperes. High precision type, expensive to manufacture. Uniform scale. Little effect on meter by external magnetic fields	D.C. and a.c. of power frequencies (100 Hz unless manufactured specially for a particular frequency). Reads true r.m.s. Needs larger currents for full scale deflection, possibly 100 mA. Non-linear scale, often a section near zero which is uncalibrated. Cheap and robust

USE OF THE WATTMETER

The power in a single phase circuit = $VI \cos \phi$ watts.

The wattmeter therefore needs both current and voltage elements to measure power. In the dynamometer wattmeter the circuit current flows in a pair of fixed coils so setting up a magnetic field. A moving coil carrying a current proportional to the circuit voltage is situated in this field. The deflection of the moving coil is a function of the strength of the magnetic field due to the circuit current in the fixed coils, the current in the voltage coil which is proportional to the circuit voltage and the phase angle between these two.

The terminal arrangement of a typical laboratory wattmeter is shown in *Figure 8.18*.

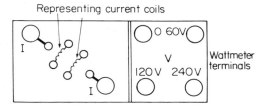

Figure 8.18

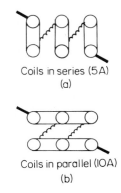

Coils in series (5A)
(a)

Coils in parallel (IOA)
(b)

Figure 8.19

There are two current coils shown but in many instruments these are each further subdivided into two making four coils in all. Each coil is capable of carrying 5 A typically. For circuit currents up to 5 A the two coils are connected in series. For currents up to 10 A they are connected in parallel. With four coils the parallel connection gives a current limit of 20 A. Movable links are provided to effect the change. The two connections are shown in *Figure 8.19*. Different voltage ranges are also provided, typically from zero to 60 V, 120 V and 240 V. As with a voltmeter, these are obtained by using different series resistors.

Before connecting a wattmeter, the circuit in which the power is to be measured should be set up including a voltmeter and an ammeter, the circuit energised and the meter readings noted. The wattmeter may now be connected using the appropriate ranges.

For a circuit drawing less than 5 A at a voltage between 120 and 240 V, the schematic and actual arrangements are shown in *Figure 8.20*. If the wattmeter pointer drives backwards from the zero it is necessary

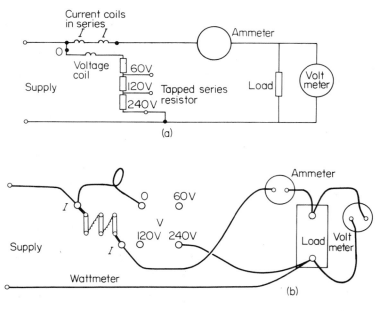

Figure 8.20

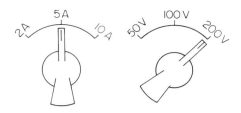

Figure 8.21

Voltage range	Current	Power
60 V	5 A	Face value
120 V	5 A	X 2
240 V	5 A	X 4
60 V	10 A	X 2
120 V	10 A	X 4
240 V	10 A	X 8

Figure 8.22

to reverse the connections to either the current coils or the voltage coil. Sometimes a reversing switch is provided for this purpose. As an alternative to movable links range switches may be provided.

Somewhere on the wattmeter casing there will be a plate giving the scale factors. Since the wattmeter has only one scale, adjustment of the current and voltage ranges alters the range of the instrument as in the case of the multimeter. A typical plate is shown in *Figure 8.22*.

With the wattmeter connected as in *Figure 8.20*, the indication on the meter face must be multiplied by 4. When recording power, both the actual wattmeter indication and the scale factor should be noted to avoid speculation at a later time as to whether all the information recorded is correct.

PROBLEMS FOR SECTION 8

(1) What is the function of a shunt when used in association with a moving-coil instrument? Draw a simple sketch showing the arrangement of meter and shunt.

(4) What is the function of a multiplier when used in association with a moving-coil instrument? Draw a simple sketch showing the arrangement of meter and multiplier.

(5) Calculate the value of a shunt to be used to enable 0–15 μA moving-coil ammeter with coil resistance of 5 Ω to be used in a circuit with maximum current 0.15 A.

(6) Calculate the value of a multiplier to enable the meter in Problem 5 to be used as a voltmeter with full-scale deflection at 15 V.

(7) Why is a rectifier necessary when a moving-coil instrument is to be used in an a.c. circuit?

(8) A moving-coil rectifier instrument is used as an ammeter in a circuit in which the current is non-sinusoidal. Explain why the meter indication cannot be trusted.

(9) Why is damping necessary in a voltmeter?

(10) How is damping achieved in a moving coil meter?

(11) What is the function of the hair springs in a moving-iron meter?

(12) A wattmeter with a scale plate as in *Figure 8.22* is used to measure the power in an a.c. circuit. The indication is 53 W. The current coils are connected in parallel and the voltage range is 240 V. What is the value of the power in the circuit?

9 Planned maintenance

Aims: At the end of this section you should be able to:
State the advantages of preventive maintenance.
Explain the need to plan maintenance.
Describe the testing of cables, motors, transformers and switchgear.
State the dangers when isolating and earthing electrical circuits.
State the reasons for, and the methods of, earthing.
Explain the need for locks and warning notices.
Describe and give examples of accident situations.

PREVENTIVE MAINTENANCE

There are two approaches to the maintenance of plant:
1. Allow it to run until it breaks down and then carry out repairs.
2. Regularly test and inspect it, replacing worn parts, oiling and greasing, revarnishing windings etc., to attempt to prevent breakdowns as far as possible.

The approach depends in part on the process being carried out. A breakdown may result in a period of non-production with loss of revenue from the plant. It may cause considerable damage to associated plant. The breakdown of an oil pump providing lubrication for a gear box on a large piece of equipment could cause the gear box to be wrecked. The failure of a circuit breaker attempting to open to clear a fault could cause severe burning and disruption of the system. On the other hand the failure of a vacuum cleaner motor in the home would probably be merely inconvenient.

Preventive maintenance does not generally prevent breakdowns altogether but minimises their incidence. Such work can often be carried out during periods of light loading on the factory or during the holidays of the production staff rather than at the most inconvenient time, which is when breakdowns always seem to occur.

Routine maintenance allows the work force to be employed steadily for periods of the year instead of day and night occasionally during periods of breakdown.

The frequency of testing and inspection will depend on one or more of the following factors.

The type of equipment and the environment in which it works

A fully submersible pump with an electric drive working in an acidic fluid, running intermittently during 24 hours each day, would be more likely to fail in service than a small motor driven fan, running one shift per day, used to ventilate a small workshop in which light assembly work is carried out. The interval between checks on the submersible pump would need to be shorter than for the fan.

Plant history

Where a plant has a record of failure at regular intervals, or when inspections formerly took place worn parts were generally found, the periods between inspections would be shortened to allow parts to be replaced before conditions can be created which would lead to breakdown.

Plant running hours

Electronic apparatus using printed circuit boards might be allowed initially to run to failure. In the light of the experience gained, future board or unit changes might be made after less than the mean time to failure. Where electric motor and generator brushgear is concerned, experience will show how many running hours results in brushes wearing down to unsafe lengths. Having established this period, brushes will be changed after the plant has run say 20 per cent less than the critical time.

Availability of plant for testing

Certain plant is only available for testing and maintenance after fixed intervals so that all work other than breakdown maintenance must be carried out at this time. Steam boilers, for example, may run for 14 or 26 months, according to type, between statutory inspections. This may fix the period between inspections for non-duplicated auxiliary equipment.

The plant maintenance engineer generally plans at least one year ahead when the running programme for the factory is known. The availability of spare parts has to be checked and orders placed where thought necessary.

Inspection and tests are detailed under 'essential' and 'optional' headings, and a length of time and number of staff allocated to each job. The number of maintenance staff available will determine how many jobs can be carried out simultaneously. The progress has to be such that the whole plant can be returned to service when required for production. Failure to do so results in financial loss and the amount of this loss is one measure of the effectiveness of the maintenance department. Some of the optional jobs may have to be held over until another time.

THE 'MEGGER' INSULATION TESTER (EVERSHED AND VIGNOLES LTD.)

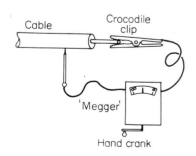

Figure 9.1

In Chapter 8 the ohmmeter was described and this is suitable for the measurement of the resistance of components. For cable, switchgear, and motor insulation resistance testing, a device which can measure millions of ohms whilst stressing the material with hundreds or even thousands of volts is required. The main limitation of the ohmmeter for testing insulating materials is the applied voltage. Consider a fault on an electric motor which allows a conductor to touch the casing. The resistance between the conductor and case with even the slightest amount of dirt or grease at the point of contact could be several thousand ohms when measured with the ohmmeter with its battery source. However, with normal working voltage applied to the conductor electrical breakdown of the dirt would occur causing the resistance to fall to a very low value.

The insulation testing meter often used is the 'Megger' and it comprises a hand-cranked d.c. generator and an ohmmeter type movement. Models are available at 250 V, 500 V, 1000 V and 2500 V.

Figure 9.1 shows the 'Megger' being used to measure the insulation resistance between the core and sheath of a concentric cable. The leads are connected as shown, often using crocodile clips, and the handle turned increasingly rapidly until it appears to turn more easily. The internal clutch is then slipping and this occurs at the speed required to generate the voltage of the particular test set. After a short interval depending on the capacitance of the device being tested, the pointer comes to rest on the scale when the insulation resistance may be read.

EQUIPMENT TESTING

There are many tests which are carried out after installation of new plant and during commissioning. These are very specialised and not dealt with here. The following tests and inspections are carried out on plant during periodic checks or after a breakdown when its fitness for return to service is being considered.

Cable tests

The cable must first be disconnected from the supply. In the case of a cable feeding a piece of equipment, this will involve opening the circuit breaker and locking it in the open or isolated position. To test the cable alone, or as in the next section the motor or equipment alone, the cable connections to the equipment must be unbolted and gently eased away so that there is no contact.

In the case of a feeder which can be supplied from both ends, the circuit breakers at both ends must be opened and isolated.

Electrical tests are then made using the 'Megger'. For circuits operating normally at 415/240 V, a 500 V 'Megger' is generally used. For higher voltage circuits the higher voltage test sets are used.

On single-phase circuits tests are carried out between the two cores and between each core and earth. On three-phase circuits the tests are carried out between each pair of conductors; Red to Yellow, Yellow to Blue, Blue to Red; and from each of the cores to the earthed sheath. No absolute figures can be quoted for the insulation resistance since this will depend on the length and type of cable, but the readings should generally be hundreds of megohms (1 megohm = 1 million ohms).

Visual checks should be made at the same time of the cable ends. Lead-covered paper-insulated cables have their ends terminated in compound-filled boxes. There should be no oil seeping from the box and all should appear clean and in good order. Plastic-insulated cables are terminated in a screwed gland and it is necessary to generally check for cleanliness and appearance.

Motor tests

The supply is isolated and the motor disconnected from its supply cable at the terminal box.

On a single-phase motor the insulation resistance between one motor terminal and earth is checked. In the case of a three-phase motor the windings are separated by removing the links in the terminal box. The tests are then as for a three-phase cable, testing between each winding and its neighbour and between each winding and earth. With a d.c. motor the armature and field windings can be separated and the insulation resistance of each part to earth tested.

Visual checks should be made of the coupling between the motor and its load, the windings, by removing either or both end covers, and the bearing grease or oil. A darkened grease with a bad smell often indicates trouble which could necessitate taking the motor to the workshop to fit new bearings.

Transformer tests

The input and output cables are isolated by opening the appropriate circuit breakers and locking them in this position. The cable ends are disconnected from the windings in the terminal boxes and the windings separated if there are available links. The insulation tests are then as for the motor on both primary and secondary windings separately.

Where a tap changer is fitted a visual check of the mechanism and contacts is advisable if accessible. With oil-filled transformers the cold oil level is checked and a sample of oil taken for checks on electrical breakdown value, moisture content and acidity. Filtering or replacement may be required which, if suitable plant is not available in the factory, will be undertaken by the oil supplier.

Due to the inherent reliability of transformers the full range of tests is rarely carried out, a watch on the condition of the oil being considered adequate unless trouble is suspected.

Switchgear tests   The switch is isolated from the supply and a visual check made on the linkages, contacts and any operating coils, after removing the oil tank on oil-filled gear. The switch is closed manually to check contact movement and alignment. Any operating coils are isolated electrically and the 'Megger' used to check insulation resistance between the windings and the chassis of the switch.

With the switch open, insulation tests are made between Red and Yellow, Yellow and Blue, and Blue and Red terminals on both the input and output sides. Finally insulation tests are carried out between the input and output terminals of the corresponding colours and each to earth.

With oil-filled gear a visual check for blackening of the oil is carried out. Similar tests can be carried out on the oil as for the transformer but colour and smell are usually good enough indications of condition.

ISOLATION AND EARTHING

When work is to be carried out on high and extra-high voltage equipment, isolation from all sources of danger must be carried out and to prevent the potential of that equipment rising above that of earth due to accidental contact with live metal or inductive effects, it is solidly connected to earth. Generally a Permit-to-Work is issued stating the points of isolation and earthing and the precise limits of the safe area and equipment to be worked on. If the earth connection has to be removed for testing purposes, a special Sanction-for-Test document is required.

At low and medium voltages isolation from the supply is sufficient without earthing. Here again some form of Permit-to-Work is desirable.

For isolation at 415/240 V an isolator or switch fuse is often used. The schematic arrangement is shown in *Figure 9.2a* while a typical enclosure together with the method of locking is shown in *Figure 9.2b*.

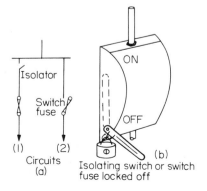

Figure 9.2

Before work is commenced on a circuit the isolator or switch fuse is opened and locked in this position. The key is held either by the workman involved or in a lockout box, a key to which he holds. A lockout box has several different locks and keys so that several interested parties can each hold a key. They all have to come together before the box can be opened and the circuit re-energised.

On h.v. or e.h.v. circuits up to 33 kV isolation is often achieved using withdrawable type switchgear.

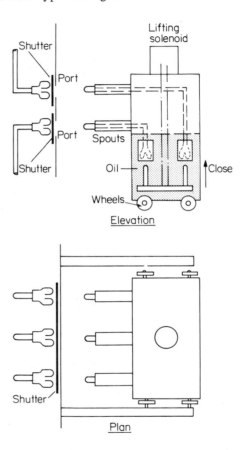

Figure 9.3

Figure 9.3 shows two views of a bulk oil circuit breaker, for use up to 33 kV, with horizontal isolation. The circuit breaker is mounted on wheels which run on tracks. The contacts are of the rose and poker type. In order to energise the circuit, the circuit breaker truck is pushed into its cubicle. The shutters are lifted at the last possible moment by small wheels on the side of the breaker engaging lifting levers attached to them. The spouts enter the ports and make contact with the female rose contacts. The truck is locked in position. The circuit breaker may now be operated normally.

To isolate the circuit breaker it must be in the open position when it may be withdrawn from its cubicle. The metal shutters drop closing off the ports so that neither bus-bar or circuit roses may be touched. The shutters are locked in this position if work is to be done on the circuit breaker.

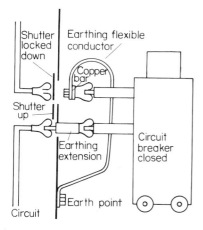

Figure 9.4

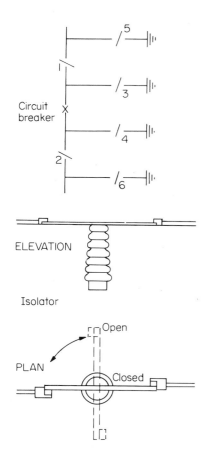

Figure 9.5

If work is to be done on the circuit cable or equipment an extension piece is fitted to each of the circuit spouts as shown in *Figure 9.4*.

A copper bar is screwed to the top (bus-bar) spouts and is solidly earthed. The circuit breaker is pushed into the cubicle when only the bottom shutters open allowing the earthing extension pieces to make contact with the circuit roses. The circuit breaker is closed so connecting the circuit to earth. If work is to be done on the bus-bars these must be disconnected from the supply transformer or generator and the positions of the extension pieces and copper earthing bar exchanged. The circuit breaker will now earth the bus-bars when racked in and closed.

Vertical isolation is also common. This is virtually the same arrangement except that the circuit breaker is raised or lowered to give contact with the bus-bars and circuit, the spouts being arranged across the top of the tank.

In open air substations at 132 kV and 400 kV, isolators are used as shown in *Figure 9.5*.

To work on the circuit breaker it is first opened. Isolators 1 and 2 are opened. Isolators 3 and 4 are closed connecting the conductors to earth. Further isolators 5 and 5 are provided to earth the incoming and outgoing lines if these are de-energised by opening the circuit breakers at the far ends of the lines.

The main danger involved is that a live circuit may be directly connected to earth. On withdrawable gear this may be brought about by connecting the extension pieces and earthing bar to the wrong sets of spouts due to a misunderstanding or carelessness. Where fixed isolators are used as in *Figure 9.5*, an incorrect sequence of operations, closing the earthing isolators before opening the circuit breaker and its isolators 1 and 2, puts earths on to a live line.

Further suppose that the circuit has been correctly earthed and that work in the cubicle itself or on one of the earthing isolators has to be done. The circuit breaker truck may have to be withdrawn or the isolator in question opened. To enable this work to be done, temporary earths have to be fitted. These are lengths of copper braid with a G clamp fitted to one end. The G clamp is screwed directly on to the bus-bar or circuit conductor whilst the other end of the braid is connected to earth. Inadvertently attempting to connect the G clamp to the wrong part of the circuit which is still alive will result in an explosion with possibly fatal results to the person concerned. To avoid this possibility a number of live-line detectors have been developed for use up to 33 kV. These sometimes indicate live conditions when the metal being tested is in fact isolated but is being charged inductively from an adjacent circuit. For this reason these testers tend to be distrusted.

Accidents can of course occur due to incompetence or forgetfulness but where a proper system of safe working has been instituted and is enforced, these should be very rare.

Accidents occur at medium voltage in the main where electricians and others are allowed to perform their own isolations. A few examples will illustrate the point.

A fitter decides to make a small adjustment to a contactor without isolating the circuit since it is easier to test the circuit whilst alive. A flashover occurs when a screwdriver touches two live terminals.

Fuses are withdrawn from a fuse-board which constitutes an isolation. Another person restores the supply unexpectedly by borrowing fuses from another way in the fuse-board.

Painters decide to work in the path of a travelling gantry crane without having it locked off.

Work is carried out on electric heaters which happen to be cold and therefore thought to be switched off, but are in fact controlled by a thermostat which operates whilst the work is going on.

A safe system of working would prevent all such occurrences.

REQUIREMENT FOR LOCKS AND WARNING NOTICES

When a circuit has been correctly isolated and earthed, the shutters are locked in position or on outdoor gear the isolator handles are locked off, the circuit breaker operating mechanism is locked off and if earthing is through a circuit breaker, the mechanical tripping device is made inoperative and locked in that position. For medium voltage work, switch fuse handles are locked in the off position.

The locks are to prevent anyone from tampering with the isolation or removing earths by oversight or in order to do unauthorised testing which may hazard someone else working on the equipment. Warning notices quoting any Permit-to-Work number and possibly the name of the issuing officer are placed at all control positions, at the circuit breaker and at the equipment upon which it is safe to work. If relevant, an area may be roped off within which it is safe to work. The rope is marked with flags and notices at regular intervals. The Permit-to-Work is issued as stated previously.

PROBLEMS FOR SECTION 9

(1) What are the advantages of planned versus breakdown maintenance?

(2) What factors govern the choice of time intervals between inspections and maintenance work on plant?

(3) Why is a battery operated ohmmeter inadequate for cable insulation testing?

(4) How does one know when the correct output voltage from a 'Megger' tester has been attained?

(5) What visual checks can be carried out advantageously on (a) an electric motor (b) cable terminations.

(6) What three tests are carried out on samples of transformer oil during routine inspections?

(7) What is the function of (a) a Permit-to-Work (b) a lock-out box?

(8) What is meant by (a) horizontal isolation (b) vertical isolation?

(9) Describe one situation concerning electrical apparatus which could lead to an accident.

(10) What is the function of a Warning Notice when posted in conjunction with a Permit-to-Work?

10 Passive and active components in electric circuits

Aims: At the end of this section you should be able to:

Describe the construction of various types of resistor.

Discuss temperature coefficient, frequency of operation and power dissipation with reference to various types of resistor.

Use the resistor colour code.

Describe the properties of non-electrolytic and electrolytic capacitors.

Discuss the various materials used in the construction of inductor and transformer cores.

Use simple test equipment to determine the correct terminal locations of active devices and the failure of passive devices.

TYPES OF RESISTOR

Moulded carbon. These are made from carbon dust, refractory material and resin, compressed to the correct shape and then heat cured. The final resistance is largely a function of the ratio of carbon dust to refractory material. The connections are made either by spraying each end with metal and then soldering tags to this deposited metal or by using inserted ends as shown in *Figure 10.1a.*

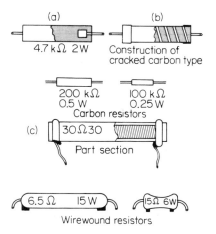

Figure 10.1

Carbon film. A glass tube is coated with a thin film of carbon which is then heated to fix the material and stabilise the value of resistance.

Cracked carbon. Carbon is deposited at high temperature on ceramic rod. Spiral grooves are cut in the carbon to give the final resistance value. Varnish may be applied overall as protection (*Figure 10.1b*).

Wirewound. These are made by winding nickel-chromium or copper-nickel wire on a ceramic former. The resistor formed may be coated in vitreous enamel and fired to give protection (*Figure 10.1c*).

Metal film.   Nickel-chromium or gold-platinum is evaporated and deposited inside a ceramic tube. The connections are made by silver plating the ends.

Oxide film.   Tin-antimony oxides are deposited on a ceramic tube forming a glass-like surface.

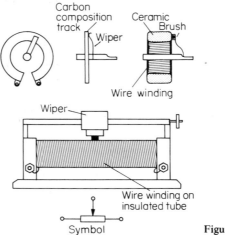

Figure 10.2

Figure 10.2 shows the arrangement of the carbon track and wire-wound variable resistors. The carbon type is almost invariably used as a low rated potentiometer whereas the wirewound types may be used both as potentiometers and as series resistors since their current carrying capability is greater.

RESISTOR CHARACTERISTICS

Temperature coefficient of resistance

When a resistor suffers a temperature change its resistance changes. Each material has a temperature coefficient of resistance which is generally expressed as the change in resistance which occurs when a 1 ohm resistor made of the material is heated from $0°C$ to $1°C$.

For copper this value is 0.0042

This means that a piece of copper with a resistance of 1 Ω at $0°C$ has a resistance of 1.0042 Ω at $1°C$ and 1.042 Ω at $10°C$ etc.

Approximate values for the various materials used in resistor construction are:

Carbon ± 0.001
Wirewound: Nickel-chromium +0.00007   Copper-nickel +0.00002
Metal film: dependent on the material used, value lies between 0.0002 and 0.0006
Oxide coating: ± 0.0005.

The largest change can be seen to occur in carbon.

Frequency response

Carbon resistors up to about 10 000 Ω in value act as pure resistors at all frequencies up into the megahertz range. At extremely high frequencies their capacitance has to be taken into account. Inductance rarely needs to be considered except where the grooving on the cracked carbon type creates what are, in effect, turns. This has to be taken into account at very high frequencies. Film and oxide types are non-capacitive and non-inductive up into the hundreds of megahertz range.

Wirewound resistors have considerable inductance and this has to be considered when using frequencies outside the power range (50–400 Hz). Double windings can be used to minimise the effect.

Power dissipation   All the carbon types are made generally up to 2 W rating whilst some are available up to 5 W.

The metal film type has similar ratings.

The oxide film construction allows up to 6 W dissipation.

Wirewound resistors may be made almost any size and ratings of several hundred watts are common.

Temperature limits   Above a specified limit permanent damage to the component may occur. The following top temperatures are generally adhered to.
Carbon 110°C
Cracked carbon 150°C
Wire 300°C except where high precision is required when less than 100°C is recommended to minimise the temperature coefficient effect.
Oxide film 300 °C

THE RESISTOR COLOUR CODE

It is desirable that circuit components in electronic circuits shall be marked with their values or a recognition symbol to facilitate assembly work and replacement of components during repairs.

Resistors are generally colour coded with the exception of wirewound types which often have the actual values marked (see *Figure 10.1c*).

The colour code splits the value into four component parts:

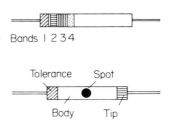

Figure 10.3

The colour of the first band or body of the resistor indicates the first
   figure.
The second band or tip indicates the second figure.
The third band or spot indicates the multiplier (the number of noughts
   to be added).
The fourth band or left hand tip when viewing the spot indicates the
   tolerance.
Where no fourth band or left hand tip colour is given the tolerance is
assumed to be 20%, i.e. that the actual value is within 20% of the value
indicated by the code markings.

Colour code

Colour	First band or body First figure	Second band or tip Second figure	Third band or spot Times	Fourth band or left hand tip Tolerance
Black	0	0	1	—
Brown	1	1	10	1%
Red	2	2	100	2%
Orange	3	3	1 000	3%
Yellow	4	4	10 000	4%
Green	5	5	100 000	—
Blue	6	6	1 million	—
Violet	7	7	10 million	—
Grey	8	8	—	—
White	9	9	—	—
Gold			0.1	5%
Silver			0.01	10%
No colour				20%

Example (1).   Determine the values and tolerances of the resistors identified as follows.

Or band	Body 1	Tip 2	Dot 3	L.H. tip 4
(a)	Yellow	Violet	Black	—
(b)	Yellow	Violet	Brown	—
(c)	Yellow	Violet	Green	—
(d)	Grey	Red	Black	Red

(a) = 47 with no zeros = 47 Ω. No entry in the fourth column indicates a tolerance of ± 20%.
(b) = 47 × 10 = 470 Ω. Tolerance as in (a)
(c) = 47 × 100 000 = 4.7 MΩ. Tolerance as in (a)
(d) = 82 Ω with tolerance ± 2%

A 2% tolerance on a 100 Ω nominal resistor means that the actual value of any resistor with those colours will lie between 100 + 2% = 102 Ω and 100 − 2% = 98 Ω.

Example (2).   What values of resistance and tolerance where relevant are indicated by the following colours?

Band	1	2	3	4
(a)	White	Brown	Yellow	Brown
(b)	Orange	White	Yellow	Yellow
(c)	Brown	Black	Black	—
(d)	Yellow	Violet	Red	Gold
(e)	Yellow	Violet	Silver	Red

Example (3).   What colour markings would indicate the following values of resistance:
(a) 25 Ω (b) 57 Ω (c) 150 Ω (d) 58 000 Ω?

CAPACITORS

Two conducting plates separated by an insulating material or dielectric constitute a capacitor which is able to store electric charge as excess electrons on one plate and a deficiency on the other.

Figure 10.4 shows some arrangements of conducting surfaces and dielectrics. The paper insulated capacitor comprises two aluminium foils separated by very thin paper soaked in paraffin wax; the mica capacitor, foils separated by mica sheets and the ceramic capacitor, layers of silver plating with a ceramic tube as dielectric. There are other types using different conductor and dielectric materials.

A simple trimming capacitor used for fine adjustment of circuit capacitance is also shown in which variation is achieved by altering the gap between the plates using an insulated screw.

The capacitance of a capacitor is a function of the plate area and the distance between them. The larger the plate area the larger is the capacitance. The smaller the distance between the plates, the larger is the capacitance. However, making the dielectric thinner reduces the value of voltage at which breakdown occurs so that this is one of the design limitations. In addition, no dielectric is perfect and some current flows through it causing a power loss in the dielectric. This loss is a function of the thickness and of the type of insulation used.

The thickness of the dielectric is reduced to almost molecular thickness, of the order of 10^{-7} metres, in the electrolytic capacitor. Aluminium foil is chemically treated on one side so as to form aluminium oxide which becomes the dielectric in the completed capacitor. An untreated foil and paper soaked in a conducting liquid, the electrolyte, are interleaved as in the paper insulated capacitor in *Figure 10.5*.

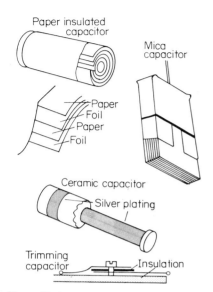

Figure 10.4

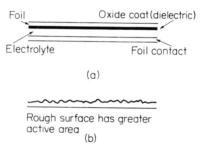

(a)

Rough surface has greater active area
(b)

Figure 10.5

The untreated foil is merely a contact for the electrolyte and the effective distance between conducting surfaces is the thickness of the oxide.

The electrolytic capacitor of this type is much smaller than the paper insulated variety for the same capacitance. Further reductions in size are obtained by etching the surface of the foils with acid or by forming them on a close mesh of aluminium (*Figure 10.5b*). Developments using tantalum foil have created further reductions in size for circuits with low working voltage.

The electrolytic capacitor must be used only in circuits with permanent d.c. bias since the action of alternating voltages is to destroy the oxide coating. They are marked for polarity, working voltages and capacitance. Non-electrolytic capacitors are suitable for a.c. or d.c. working. The table shows the characteristics and some uses of the common types of capacitor.

Figure 10.6

Type	Capacitance, typical values	Maximum voltage	Remarks
Paper	250 pF–10 μF	150 kV	Cheap, used in circuits where losses are not important.
Mica	25 pF–0.25 μF	2000 V	High quality, used in low loss circuits.
Ceramic	0.5 pF–0.01 μF	500 V	High quality, used in low loss precision circuits.
Electrolytic	1 μF–1000 μF	600 V	Used where large capacitance is needed, circuit normally contains a d.c. component of voltage. Must be connected with correct polarity. Greater loss than the paper type.
Trimmer	2 pF–150 pF	350 V	Fine variation in capacitance.

CORES FOR CHOKES AND TRANSFORMERS

When a conducting material is situated in a region of changing magnetic flux a voltage is induced in that material and a current flows. By Lenz's law the direction of this current is such as to create an opposing magnetic flux. *Figure 10.7* shows a single turn wound on a conducting core. The current in the winding is increasing. The core flux is therefore increasing and voltages are induced in (a) the current-carrying coil itself, this voltage restricting the rate of rise of the current, (b) the core material, this voltage driving what are called eddy currents which heat the core since it has resistance, (c) any other coil or conducting material in close proximity to the core, this being the basis of the transformer.

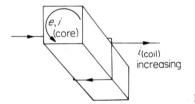

Figure 10.7

The induced voltage in case (a) is a self-induced voltage, the coil possessing self-inductance whilst in cases (b) and (c) we have mutually induced voltages, there being mutual inductance between the coil and core and any other conductor in the vicinity.

From Faraday we know that the induced voltage

$$e = \frac{\text{Change in flux linkages}}{\text{Time taken for the change}}$$

Using symbols: $e = \dfrac{\Delta \Phi}{\Delta t}$ the Greek Δ (delta) being read as 'the change in'.

Considering the core as a single turn, when an alternating current flows in the coil in *Figure 10.7*, the time interval for a current to change from zero to a maximum or from a maximum to zero etc., with the corresponding changes in core flux, is a function of frequency. The greater the frequency the smaller is the time interval for the change.

Hence, the time interval $\Delta t \propto \dfrac{1}{f}$

So that $e \propto \dfrac{\Delta\Phi}{\dfrac{1}{f}}$ $e \propto \Delta\Phi f$ volts.

Now since $i = \dfrac{e}{R}$

and power = ei watts, power = $e \times \dfrac{e}{R} = \dfrac{e^2}{R}$ watts.

Where R = resistance offered to the current flowing by the core material.

The eddy current power loss $\propto \dfrac{(\Delta\Phi f)^2}{R}$

In addition, when ferrous cores are used there is a hysteresis loss which is due to the reversal of the magnetic field in the core material. Energy is required to orientate very small groups of atoms called Domains so as to create a magnetic field in one direction during a positive-going half-cycle and then to re-orientate them with opposite polarity during the negative-going half-cycle.

The power loss is a function of the maximum working flux density; how many of the domains need to be orientated to give the required field strength; and of the number of times this is carried out per second; the frequency.

Hysteresis power loss $\propto \Delta B \times f$ watts.

MINIMISATION OF LOSSES

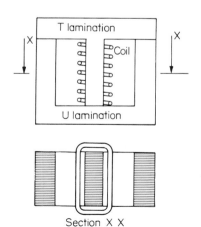

Figure 10.8

In electrical machines working at frequencies up to a few kilohertz, high fluxes are required and consequently steel with high working flux density is used so as to minimise the cross-sectional area required.

The eddy current losses in the core are reduced by increasing its resistance. This is achieved by the use of laminations. A lamination is a very thin sheet of steel, less than 0.4 mm thick, carefully cleaned and varnished or anodised on one side. Many laminations are pressed together to form the required cross section of the core. An alternative method uses thin sheets of paper between the laminations.

Figure 10.8 shows a simple arrangement. Eddy currents can now no longer flow as shown in *Figure 10.7* but are constrained to very small loops within each of the insulated laminations. Adding between 0.5 and 4% of silicon to the steel further increases the resistance at the expense of the magnetic properties.

Laminations of nickel iron at thicknesses of less than 0.1 mm may be used at frequencies up to about 10 MHz. For frequencies higher than this, since the eddy current losses are proportional to the frequency squared, these may be considerable with even the thinnest laminations so that dust or ferrite cores are used up to about 150 MHz in communications equipment. Dust cores comprise iron, nickel-iron or molybdenum particles individually covered with an insulating material

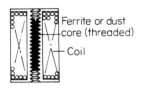

Figure 10.9

and then held together in a resinous binder. Ferrites have very high resistivities and are free from eddy current losses at all but the very highest frequencies, but since they saturate at low flux densities a larger cross-sectional area is required than when using steel.

Hysteresis losses in power equipment are minimised by the use of soft cold-rolled grain oriented steel which is very much easier to magnetize in the direction of rolling than the non-oriented variety.

In ultra-high frequency applications air cores may have to be resorted to since air suffers no hysteresis and eddy current losses.

Figure 10.9 shows a small coil with a ferrite or dust core which is threaded. It may be screwed into or out of the coil to adjust the inductance value.

FAILURE OF PASSIVE COMPONENTS

When a passive component fails its input impedance changes and tests to determine this impedance are used to decide on the nature of the failure. We will consider the three types of component in turn.

Resistors

The resistance of the resistor in question is measured using an ohm-meter or the voltmeter and ammeter method. The value of resistance obtained will indicate: (a) that the resistor is sound, the value measured being within the allowable tolerance band.
(b) an open circuit (partial or complete), the value being much higher than nominal.
(c) a short circuit (partial or complete), the value being much lower than nominal. For conditions (b) and (c), with wirewound resistors it may be possible to effect a repair but with other types, replacement is indicated. The general condition of the resistor should be considered, whether it has obviously been overheated for example. Before replacing the resistor the rest of the circuit should be examined to see whether it is possible to discover if it failed due to a malfunction of another component in that circuit.

Capacitors

A d.c. voltage slightly in excess of the rated value is applied to the capacitor through a low-range series-connected ammeter. The connections are the same as for the determination of resistance using the voltmeter and ammeter method. Care must be taken to see that the correct polarity is used on electrolytic capacitors. After a short initial charging period there should be virtually no current flowing in the circuit. Current flowing indicates leakage probably due to a failure of the dielectric.

Substituting an a.c. supply, current should flow in the circuit.

$I = V \times 2\pi f C$ amperes from which C can be calculated.

Zero, or very low current yielding a very low value of capacitance from the calculation indicates an open circuit within the capacitor.

Non-electrolytic capacitors are tested at their full rated voltage. Electrolytic capacitors will withstand a low voltage a.c. for a few seconds which is long enough to obtain readings of voltage and current to substitute in the equation.

Inductors

The resistance is measured using the same methods as for resistors. Using an a.c. source and a.c. voltmeter and ammeter, the impedance is measured.

On a.c. $\dfrac{V}{I} = Z\Omega$.

and $Z^2 - R^2 = X_L^2$ from which the inductive reactance X_L is calculated.

$X_L = 2\pi f L \ \Omega$.

The value of inductance so obtained is compared with the nominal value.
A high value of resistance indicates an open circuit.
A low value of inductance indicates short circuited turns within the coil.

SELECTION OF ACTIVE DEVICES USING MAKERS' DATA

Since there are many thousands of devices available, each with its own particular function and limitations it is not possible to give other than general guidance here. Manufacturers issue comprehensive catalogues with full information as to the working voltage, maximum current, power dissipation and temperature limitations together with graphs showing input and output characteristics. These must be carefully studied to determine whether a particular device will perform in a circuit as required.

Diodes.   The types of diode available are in the main:
Signal diodes
Power diodes
Zener or reference diodes.

Signal diodes must have a low capacitance to enable them to be used at high frequencies (see Chapter 8). They must have a low forward and high reverse resistance. They are used for the detection of radio and television signals and as switches in logic circuits. The germanium point contact diode for use as a radio detector can withstand about 30 V in the reverse direction and handle a current of 20–30 mA. Gold bonded germanium diodes are designed for use in high-speed switching circuits and have similar voltage capabilities to the point contact diode but can handle 200–400 mA.

Diodes for general use in telephone circuits and television receivers are available for 150 V working and up to 400 mA.

Power diodes can operate up to about 1000 V and handle currents of over 500 A. These are generally based on the p–n junction.

Zener or reference diodes have operating voltages from about 3--75 V. Care must be taken when selecting these diodes to see that the correct power dissipation is selected (see Chapter 11). A range is available generally up to 10 W dissipation with currents of 10–25 mA.

Transistors.   The range of transistors is even more formidable. Types are available from the JFET capable of handling a few milliamperes at 30–40 V to the high-power silicon transistors capable of operating at nearly 2000 V with a collector current of 2.5 A.

Thyristors.   For light industrial applications, a range is available for working up to 500 V and currents of up to 10 A. Heavier types can operate at 1000 V and up to 800 A. These require a gate current in the region of 1.5 A. At the other end of the scale, the low-power device can carry 1 A and the gate current is in the region of 100–200 mA.

TERMINAL LOCATIONS OF ACTIVE DEVICES

These may generally be determined using an ohmmeter with a low voltage internal supply which is often a single dry cell giving an e.m.f. of 1.4 V. It should be noted that when using a commercial multi-purpose meter, the negative (black) terminal when used for measuring voltages and currents becomes the positive terminal for resistance measurements so that the correct current direction in the meter movement is maintained. This should be clear from *Figure 10.10.*

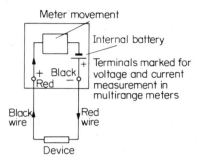

Figure 10.10

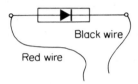

Figure 10.11

A separate battery and milliammeter may be used, a high current reading indicating a low resistance.

To avoid confusion, connect a red wire to the positive of the supply battery (black or negative terminal of a multirange instrument) and a black wire to the negative of the supply (the red terminal of a multirange instrument). The connections within this section will be referred to as red and black on this understanding.

Diodes   The ohmmeter will indicate a fairly low value of resistance when connected as shown in *Figure 10.11*. Reversing the connections will yield a near-infinity reading.

Bipolar junction transistors   *(a) pnp*   Easy current flow is from p to n.
In *Figure 10.12a*   Red on e and black on b gives low resistance.
  Red on c and black on b gives low resistance.
Reversed connections on these terminals and connections of either polarity between e and c give high resistance readings.

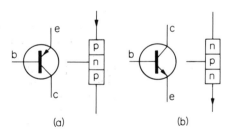

Figure 10.12

(b) npn   Again easy current flow is from p to n.
Figure 10.12b   Red on b and black on c gives low resistance.
  Red on b and black on e gives low resistance.
Reversed connections on these terminals and connections of either polarity between c and e give high resistance readings.

JFET   From gate to source or from gate to drain the resistance is nearly infinity. All that can be done here is to find a pair of terminals between which a fairly low resistance can be measured. These two terminals are source and drain. The third terminal must be the gate.

Thyristor

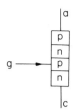

Figure 10.13

From cathode to anode with either polarity connected, a near infinity resistance reading is obtained.

With red on g and black on c a low resistance reading is obtained.

With black on g and red on c a higher resistance value is obtained. This may vary between only slightly greater and considerably greater than the previous reading according to the device tested.

These tests are only guides to connection. Reference to maker's data should be made wherever possible before connecting any of these devices into a circuit. Connections of incorrect polarity can cause permanent damage.

PROBLEMS FOR SECTION 10

(4) A moulded carbon resistor and a wirewound nickel-chromium resistor both have a resistance of 1 Ω at 0°C. What will their respective resistances be at 50°C?

(5) Why are wirewound resistors not suitable for use in circuits operating at very high frequencies?

(6) Write down the colour coding for the following resistance values:

(a) 150 Ω, 4%   (b) 58 Ω, 2%   (c) 0.58 Ω, 10%

(7) What values of resistance are indicated by the following colours?

Band	1	2	3	4
(a)	Grey	Brown	Green	Gold
(b)	Violet	Orange	Red	Silver
(c)	Brown	Yellow	Blue	Orange

(8) The capacitance of a capacitor increases as the plate area is increased. What methods are employed in capacitor manufacture to contain a very large plate area in a small space?

(9) Why is the core of a power transformer laminated?

(10) What method is adopted to reduce the eddy current losses in the cores of coils operating in the 100 MHz range?

(11) A capacitor with a nominal value of 10 μF is connected to a 100 V, 50 Hz supply when a current of 0.05 A is indicated on a series-connected ammeter. What may be deduced?

(12) A coil is tested on d.c. and the resistance is found to be 2 Ω. When connected to a 50 V, 50 Hz supply a current of 10 A flows in the coil. What is the inductance of the coil?

(13) Three, 3-terminal active devices are tested using an ohm-meter and the following readings taken using the red and black wire connections as described in the text.

(a) Red wire on terminal 1, black wire on terminal 3, R = 1 MΩ
Red wire on terminal 1, black wire on terminal 2, R = 1 MΩ
Red wire on terminal 2, black wire on terminal 3, R = 10 Ω
Red wire on terminal 3, black wire on terminal 2, R = 500 Ω

(b) Red wire on terminal 2, black wire on terminal 1, R = 100 000 Ω
Red wire on terminal 2, black wire on terminal 3, R = 10 Ω
Red wire on terminal 3, black wire on terminal 2, R = 10 Ω

(c) Red wire on terminal 1, black wire on terminal 2, R = 50 Ω
Red wire on terminal 3, black wire on terminal 2, R = 50 Ω
Red wire on terminal 1, black wire on terminal 3, R = 1000 Ω

and the same value of resistance is obtained if the polarity is reversed.

Draw a circuit diagram of each of the devices, naming them and marking the terminals.

11 D.C. power supplies for electronic apparatus

Aims: At the end of this section you should be able to:
Explain the operation of half-wave and full-wave rectifiers.
Calculate the required value, and explain the action, of a smoothing capacitor for a rectifier supplying a fixed load.
Identify the need for stabilising the output voltage of a smoothed d.c. supply.
Describe the principles and characteristics of zener diodes.
Perform calculations on, and connect a zener diode into, simple voltage stabiliser circuits.

RECTIFIER CIRCUITS

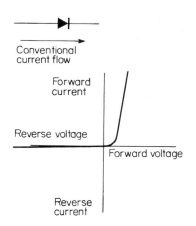

Figure 11.1   Diode characteristic (semiconductor)

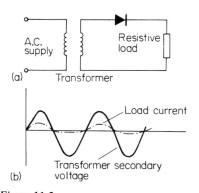

Figure 11.2

A diode is a two-terminal device which allows current to pass through it in one direction only. There are several different types, vacuum tubes, gas-filled tubes, mercury valves and those employing semiconductors. The circuit symbol is shown in *Figure 11.1*, and conventional current flows easily in the direction of the arrow.

When the diode is forward-biased current flows but when it is reverse-biased virtually no current flows unless a high enough voltage is applied to cause breakdown when damage may occur. The forward and reverse characteristics of a semiconductor diode are also shown in *Figure 11.1*, the scales of the axes depending on the material from which the diode is made.

The simplest rectifier circuit employs a single diode connected as shown in *Figure 11.2*.

An isolating transformer is used to obtain the correct voltage for the load circuit. The diode conducts for only one direction of applied voltage so that the current in a resistive load is unidirectional as shown in *Figure 11.2b*. This connection gives half-wave rectification. Using two diodes, current may be caused to flow in the load during both half-cycles and the arrangement shown in *Figure 11.3* is a full-wave rectifier.

During positive half-cycles diode D1 conducts whilst diode D2 blocks current flow so that current flows in the load through D1.

During negative half-cycles diode D2 conducts whilst diode D1 blocks the current. Current flows in the load through D2.

Only one half of the transformer winding conducts during each half-cycle and this rectifier is in effect two half-wave rectifiers back to back. An improvement in transformer usage is obtained by employing the full-wave bridge circuit as shown in *Figure 11.4*. The whole of the transformer secondary winding carries current during each half-cycle and the waveforms are again as in *Figure 11.3*.

With the rectifier connection shown in *Figure 11.2*, the load current and voltage change from zero to a peak value and back to zero again during one half-cycle of the a.c. supply whilst for those shown in *Figures 11.3* and *11.4*, the changes occur twice per cycle. The magnitude of the voltage changes is known as the ripple voltage and the number of times that it occurs per second is the ripple frequency.

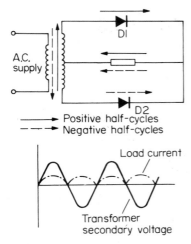

Figure 11.4

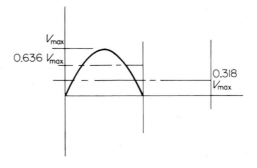

Figure 11.3

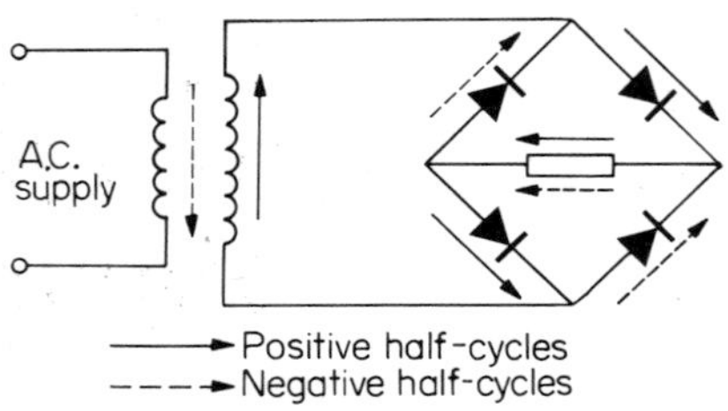

Hence: the ripple frequency $n = f$ for a half-wave rectifier.

$$= 2f \text{ for a full-wave rectifier, where}$$
$$f = \text{supply frequency.}$$

The ripple may be observed by using an oscilloscope with the input leads connected directly across the load where this is resistive. Alternatively a small value standard resistor (one with a precisely known value of resistance) may be included in the load circuit and the waveform of the voltage developed across this resistor observed. The voltage sensitivity and the sweep time or frequency can be read from the oscilloscope control panel and these readings together with the scaled size of the ripple on the screen will enable the value of the current peak to be determined, together with the ripple frequency.

Average values   The average value of a sinusoidal voltage wave over one half-cycle is 0.636 times the maximum value.

$$V_{av.} = 0.636 \, V_{max.}$$

Similarly for current: $I_{av.} = 0.636 \, I_{max.}$

For the half-wave connection, the average over the conducting half-cycle $= 0.636 \, V_{max.}$

Figure 11.5

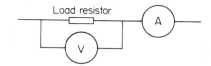

Figure 11.6

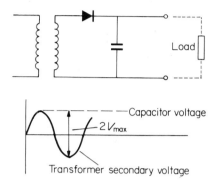

Figure 11.7

Since there is no conduction during the next half-cycle this value must be divided by 2 to give the average over the complete cycle.

Average over the complete cycle = 0.318 V_{max}.

For the full-wave rectifier, $V_{av.}$ = 0.636 V_{max}.

The average values of voltage and current for these two connections are measured by inserting moving-coil meters into the circuit as shown in *Figure 11.6*. Moving-iron meters must not be used since these indicate r.m.s. values.

Smoothing. For many applications such as audio equipment, a supply with a large amount of ripple is not acceptable since it produces hum in the output. Connecting a large capacitor in parallel with the load will reduce the ripple. Consider the diode and capacitor shown in *Figure 11.7*. The load is not connected.

During the positive half-cycle of voltage the diode conducts, charging the capacitor to the peak value of the a.c. voltage. As the supply voltage falls from its peak value the capacitor voltage is greater than that of the supply. It cannot discharge back into the supply however since the diode blocks reverse current flow. The capacitor voltage remains constant at the peak value. It is important to notice that when the transformer voltage drives negative the maximum voltage in the reverse direction across the diode is twice the peak value. This is known as the 'Peak inverse voltage' and a diode must be selected which can withstand this.

Now consider the effect of connecting the load in parallel with the fully charged capacitor. As the supply voltage falls below its maximum value the capacitor supplies current to the load and whilst losing charge its voltage falls. If the load resistance is large the current will be small and the voltage will fall only slowly and will still be at near peak value when the rectifier output voltage again goes positive. The voltage across the load therefore approaches the peak value of the supply as shown in *Figure 11.8*.

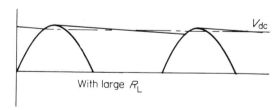

Figure 11.8

If the load resistance is small, most of the charge held by the capacitor will be lost before it is recharged during the next positive-going half-cycle. The load voltage will then be only slightly greater than the average value as indicated in *Figure 11.5*.

The capacitor is charged each time the supply voltage exceeds that remaining on it. During such periods the load is fed directly from the supply. It can be seen from *Figures 11.8* and *11.9* that the load voltage has a mean value $V_{d.c.}$ upon which is superimposed a non-sinusoidal alternating voltage which has a peak-to-peak value equal to the voltage variation.

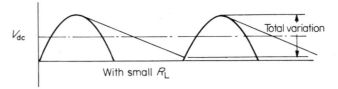

Figure 11.9

The effect of lowering the load resistance and so increasing the load current is to increase the ripple which reduces the value of $V_{d.c.}$:

The ripple can be quoted as 'total', 'peak-to-peak' or as a percentage of the value of $V_{d.c.}$ or of $V_{max.}$. A supply with $V_{d.c.} = 100$ V with a ripple of 5% of $V_{d.c.}$ would vary from 100 V + 5% = 105 V to 100 V − 5% = 95 V. This is a total or 'peak-to-peak' variation of 10 V.

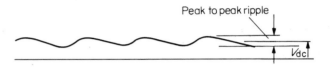

Figure 11.10

Using a full-wave rectifier reduces ripple since intervals between charges are shorter. *Figure 11.11* shows that the ripple is approximately halved by using the full-wave connection. *Figure 11.12* shows the effect on $V_{d.c.}$ of increasing load current.

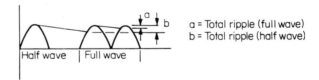

Figure 11.11

The peak voltage of the supply and hence the no-load output voltage of the rectifier is adjusted by varying the turns ratio of the isolating transformer.

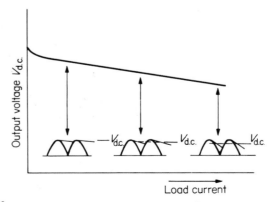

Figure 11.12

APPROXIMATE VALUE OF SMOOTHING CAPACITOR

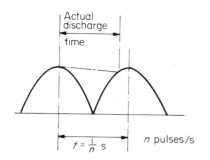

Figure 11.13

With n pulses per second, the time interval between voltage peaks $= \dfrac{1}{n}$ s.

This is very nearly equal to the capacitor discharge time for small values of ripple. Let the mean d.c. load current $= I_{\text{d.c.}}$ A.
Two identities are required.
For a capacitor, Q = Capacitance in Farads × voltage change
Where Q = charge in coulombs.
In symbols: $Q = C\Delta V$ coulombs.
Also $Q = It$ where I = current in amperes and t = time in seconds.
Now, charge given to load = charge lost by capacitor between charging pulses

Hence, $I_{\text{d.c.}} \times \dfrac{1}{n} = C\Delta V$

Transposing $C = \dfrac{I_{\text{d.c.}}}{n} \times \dfrac{1}{\Delta V}$

Example (1). A load is to be supplied with 50 mA at a mean voltage of 285 V using a full-wave rectifier. The peak value of the a.c. supply is 300 V and the frequency 50 Hz. Calculate the approximate size of a suitable smoothing capacitor.
At 50 Hz with a full-wave rectifier, the time interval between

successive peaks $= \dfrac{1}{2f}$ s $= \dfrac{1}{100} = 0.01$ s.

Discharge during this period $= I_{\text{d.c.}} \times$ time
$$= 50 \times 10^{-3} \times 0.01 = 0.5 \times 10^{-3} \text{ C}$$

Mean d.c. voltage = 285 V. Peak voltage = 300 V. Change in voltage from mean to maximum = 15 V. By symmetry, the change from mean to minimum voltage is also 15 V.
Total (peak-to-peak) voltage = 30 V (see *Figure 11.8*).

Therefore $C \times 30 = 0.5 \times 10^{-3}$ $\qquad C = \dfrac{0.5 \times 10^{-3}}{30}$

$$= 16.67 \times 10^{-6} \text{ F}$$
$$(16.67 \ \mu\text{F}).$$

Example (2). Estimate the size of a suitable smoothing capacitor for the following duty when a full-wave rectifier is used.
$V_{\text{d.c.}} = 400$ V, Ripple = 3% of $V_{\text{d.c.}}$, Load resistance = 5000 Ω, Frequency = 50 Hz.

Voltage change is from 400 V + 3% to 400 V − 3%
from 412 V to 388 V   Hence $\Delta V = 24$ V

Load current $I_{\text{d.c.}} = \dfrac{V_{\text{d.c.}}}{R_{\text{L}}} = \dfrac{400}{5000} = 0.08$ A

$C = \dfrac{I_{\text{d.c.}}}{n} \times \dfrac{1}{\Delta V} = \dfrac{0.08 \times 1}{2 \times 50 \times 24} = 33.3 \ \mu\text{F}.$

Example (3).   Estimate the size of a smoothing capacitor to give a mean value of d.c. voltage of 240 V when the peak supply voltage is 250 V using a half-wave rectifier connected to a 50 Hz supply (a) with a load current of 5 mA
(b) with a load current of 100 mA.

Example (4).   A 50 μF capacitor is used for smoothing the output from a full-wave rectifier connected to a 50 Hz supply. If $V_{\text{d.c.}}$ = 150 V estimate the total voltage variation and the ripple expressed as a percentage of $V_{\text{d.c.}}$ for R_L = 3000 Ω.

The suitability of these smoothing capacitors for the particular duty may be determined by viewing the load voltage waveforms using an oscilloscope as described in the section on 'Ripple'.

IMPROVED SMOOTHING METHODS

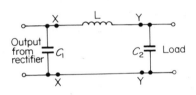

Figure 11.14

Even when a very large capacitor is used, the ripple can still be excessive for many applications. An extremely smooth output can be achieved by adding inductance and a further capacitance to the circuit as shown in *Figure 11.14*.

The output voltage from the rectifier is applied to capacitor C_1 and the voltage at XX in *Figure 11.14* is as shown in *Figure 11.10*. The load current flows through the inductor L in which voltages are induced during current changes.

The induced voltage $e = L \, di/dt$ volts, where L = inductance of the inductor in henries.

By Lenz's law, the induced voltages will oppose those causing the current changes. Consider the ripple voltage. As this increases and the load current increases, the induced e.m.f. in the inductor will oppose the increase. As the ripple voltage falls and the load current tends to fall, the induced e.m.f. will assist the current. The voltage at YY and the current in the inductor are therefore almost ripple free. A further capacitor C_2 completes the process.

The smoothing inductor can be expensive and the cost of smoothing may be reduced by replacing the inductor with a resistor which is often of several hundred ohms in value. The capacitor sizes are increased at the same time.

Again consider the ripple voltage. As the voltage increases the current in the resistor R increases.

Now, $V_{YY} = V_{XX} - IR$ (*Figure 11.15*)

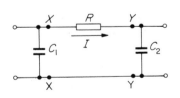

Figure 11.15

V_{YY} is somewhat lower than V_{XX}. As the ripple voltage falls to its minimum value it approaches that at V_{YY} which is held on the capacitor C_2. The current I falls to a very low value so that the volt drop IR is very small. In this manner V_{YY} is kept very nearly constant. The output voltage is lower than with inductor smoothing due to the resistive volt drop, and there is a loss of efficiency due to the power loss in the resistor.

VOLTAGE STABILISATION

The change in voltage which occurs due to a change in load can have undesirable effects. Consider as an example a simple cassette tape player. The speed of the drive motor is a function of the voltage applied to its terminals. When playing a tape, increasing the output volume causes more current to be drawn from the power supply which, as we have already seen, reduces the output voltage. It clearly would be most unsatisfactory if every time a loud passage were played, the voltage fell and the tape travel slowed.

It is therefore necessary to stabilise the output voltage from a rectifier circuit for a number of applications. One way in which this can be achieved is by the use of a zener diode.

The zener diode. By modifying the construction of a normal junction diode made from semiconducting (p–n) material, the characteristic in the reverse direction is modified from that shown in *Figure 11.1* to that shown in *Figure 11.16*. A device with such a characteristic is known as a Zener diode. It is non-conducting up to a precisely known reverse voltage, 6.8 V in *Figure 11.16*, when it suddenly becomes conducting. The result of trying to further increase the voltage is to cause it to conduct whatever current is necessary to produce voltage drops in circuit components and connections to maintain 6.8 V at its terminals even if in so doing it destroys itself by overheating.

The voltage at which conduction commences in a particular diode is specified in makers' catalogues and devices are purchased for particular applications.

Simple voltage stabilisation is obtained using a stabilising resistor as shown in *Figure 11.17*.

Suppose the output in the circuit shown is to be maintained at 12 V whilst the rectifier output is in the region of 16 V.

A 12 V zener is used and V_{DE} cannot rise above 12 V. No load is connected.

$$V_{DE} = 12 \text{ V} \qquad V_{AE} = 16 \text{ V} \qquad \text{Therefore } V_{AD} = 4 \text{ V}$$

$$I = \frac{V_{ad}}{R} = \frac{4}{2} = 2 \text{ A.}$$

The power dissipated in the zener diode = VI watts = $12 \times 2 = 24$ W so that it must be mounted on a heat sink of this rating.

Consider now connecting a load resistor of value 12 Ω to the output terminals.

$V_{DE} = 12$ V, therefore $I_L = \dfrac{12}{12} = 1$ A.

Since a current of 2 A must flow in the stabilising resistor to maintain the 4 V difference between input and zener voltages, the zener current must fall to 1 A.

For a load resistance of 6 Ω, the load current is 2 A and therefore no current will flow in the zener diode.

Should the load current increase above 2 A, the load voltage will fall below 12 V. Stabilisation is achieved for any current between 0 and 2 A with this circuit and up to this value the load on the rectifier remains constant at 2 A. The zener diode acts as a bypass for the current not required by the load.

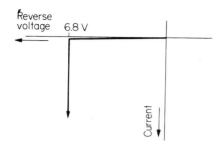

Figure 11.16

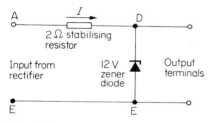

Figure 11.17

Example (5). A 10 V zener diode is used to stabilise a d.c. supply at 10 V for load currents between zero and 0.5 A. The input voltage is 15 V.

Determine (a) the value of a suitable stabilising resistor (b) the current and power dissipated in the zener diode when the load current is 0.2 A.

Volt drop across the stabilising resistor = 15 − 10 = 5 V
When a load current of 0.5 A flows, the zener diode carries no current.
Therefore 0.5 × R_s = 5 V. R_s = 10 Ω.
With 0.2 A in the load, the zener diode current = 0.5 − 0.2 = 0.3 A.
Power dissipation = 0.3 × 10 = 3 W.

Example (6).   A 22 V supply is to be used to supply a load at a constant voltage of 18 V. An 18 V zener diode and a 10 Ω stabilising resistor are used. Calculate the current range over which the stabilisation is effective.

Volt drop in the stablising resistor = 22 − 18 = 4 V.

Maximum current in a 10 Ω resistor = $\dfrac{4}{10}$ = 0.4 A.

The current range is from zero to 0.4 A.

Example (7).   A 37 V zener diode is to be used to stabilise the voltage across a load, the minimum value of which is 20 Ω. A 5 Ω stabilising resistor is to be used. What value of output voltage is required from the rectifier?

PROBLEMS FOR SECTION 11

(8) Determine the average value of d.c. output voltage from (a) a half-wave rectifier, (b) a full-wave rectifier, when fed from a 20 V a.c. supply.

(9) What is the value of the ripple voltage and the ripple frequency delivered from the output terminals of a full-wave rectifier which is fed from a 9 V, 50 Hz supply?

(10) A smoothed rectified voltage has a mean value of 120 V and 5% ripple expressed in terms of $V_{d.c.}$. What are the maximum and minimum values of the output voltage?

(11) Estimate the size of a smoothing capacitor to give a mean value of d.c. voltage of 50 V when the peak supply voltage is 52 V using a full-wave rectifier connected to (a) 50 Hz and (b) 400 Hz mains. The load resistance = 500 Ω.

(12) Determine the value of ripple expressed as a percentage of $V_{d.c.}$ when a 100 μF capacitor is used for smoothing the output from a full-wave rectifier connected to a 400 Hz supply when $V_{d.c.}$ = 12 V and the load current = 10 mA.

(13) Under what load conditions is a zener diode connected in a stabilising circuit most likely to fail due to overheating? How is the heat produced dissipated?

(14) A constant voltage of 12 V is to be maintained across a load using a zener diode and stabilising resistor.

Determine: (a) the required input voltage for a load current of 0.1 A when using a 5 Ω stabilising resistor, (b) the maximum current which can be delivered to the load at 12 V if the input is at 18 V and the stabilising resistor has a value of 24 Ω.

For both cases determine the maximum power dissipation in the zener diode.

12 Basic transistor amplifiers

Aims: At the end of this section you should be able to:

Sketch the structure, and explain in general terms the operation of, the n-channel and p-channel junction FET.

Sketch a set of output characteristics for a junction FET.

State the three configurations in which the bipolar junction transistor and the junction FET are used.

Draw circuit diagrams of simple single stage audio-frequency amplifiers.

Superimpose the d.c. load line on the output characteristics of both types of transistor.

Describe the effect of varying bias conditions and select the optimum value of bias.

Calculate voltage, current and power gains of transistors using the d.c. load line and a.c. input resistance.

THE JUNCTION FIELD EFFECT TRANSISTOR (JFET)

The JFET is made up from p and n-type semiconducting materials in the form of a sandwich, either a layer of n-type material between two layers of p-type material (n-channel) or a layer of p-type material between two layers of n-type material (p-channel).

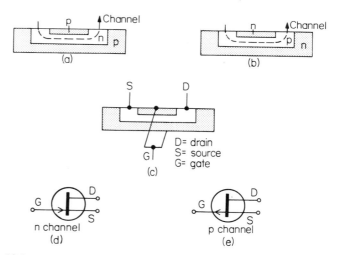

Figure 12.1

In either type current flows through the channel from the source to the drain. The two outer layers are connected electrically and form the gate through which current must flow. *Figure 12.1* shows cross-sectional views and circuit symbols. The actual thickness of the device is only a small fraction of a millimetre.

Some typical terminal arrangements are shown in *Figure 12.2*. Manufacturers' data must be consulted for their own particular terminal configurations.

OPERATION OF THE JFET

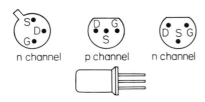

Figure 12.2

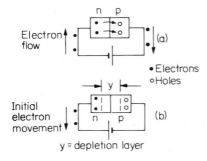

Figure 12.3

In *Figure 12.3*,   n = donor material (having excess electrons)

p = acceptor material (having two few electrons leaving *holes*)

The device illustrated is a p–n junction diode. In *Figure 12.3a* the battery source supplies electrons to the n-type material and electrons flow into the p-type material filling holes. As electrons leave the p-type material more holes are formed. The process is continuous and the diode is a good conductor. We say that in this condition it is forward biased.

When the source of e.m.f. is reversed in polarity as shown in *Figure 12.3b*, electrons are attracted away from the n-type material whilst holes are filled in the p-type material and a depletion layer is formed in which there are no free electrons or holes. This layer is in consequence a good insulator. The diode is now reverse biased.

Now consider the n-channel JFET as shown in *Figure 12.4*. Both layers of the n-type material are connected to the negative terminal of the gate supply. The n-type material of the channel is connected to the positive terminals of both supplies. The p–n junctions formed on both sides of the device are therefore reverse biased so that depletion layers are formed as shown in *Figure 12.4c*. The area of n-type material in the channel through which current can flow is restricted. The thickness of the depletion layers is a function of the gate/source voltage, V_{GS}. The

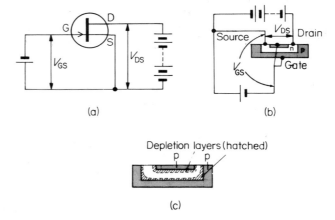

Figure 12.4

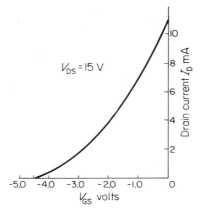

Figure 12.5

greater is V_{GS} the thicker is the depletion layer. Reducing the area of the channel increases its resistance. The relationship between the drain current flowing through the channel, I_D, and the voltage V_{GS} whilst maintaining V_{DS} constant is shown for a typical device in *Figure 12.5*.

Notice that when V_{GS} reaches -4.5 V the depletion layers are thick enough to block off the complete channel and the drain current falls to zero. This is known as the pinch-off voltage.

JFET CHARACTERISTICS

Using the circuit shown in *Figure 12.4a* together with two voltmeters and an ammeter, a set of readings may be obtained from which the characteristics shown in *Figure 12.6* have been plotted.

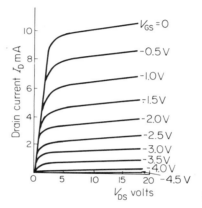

Figure 12.6

For V_{GS} at zero, V_{DS} is varied in steps over a range from zero to the maximum value specified for the particular device by the manufacturer. At each step the drain current I_D is noted.

Further readings of V_{DS} and I_D are taken for values of V_{GS} up to that value which results in zero drain current, i.e. the pinch-off value.

The curves in *Figure 12.5* and *12.6* are typical in shape but are not for any particular device.

THE JUNCTION TRANSISTOR (BIPOLAR JUNCTION TRANSISTOR BJT)

Figure 12.7 shows a schematic diagram of a junction transistor in common-base configuration together with its circuit symbol. It is made up of two diodes. One of these is the junction between n_1 and p and the other is the junction between p and n_2. The junction between n_1 and p is forward biased by battery B_1 so that electrons can flow readily from n_1 to p. The n_1 layer is termed the emitter.

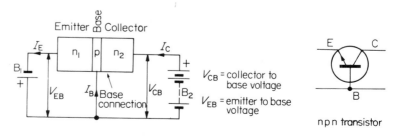

Figure 12.7

The junction diode formed between n_2 and p is reverse biased so that with battery B_1 disconnected, this junction is non-conducting.

With battery B_1 connected, electrons flow from n_1 to p and would complete the circuit back to the battery through the base connection if it were not made very difficult for this to happen. The base layer material is made extremely thin and so has a high resistance. Also, any electrons in the base layer are extremely close to the strong positive potential of battery B_2 in the n_2 layer of the transistor. Most of the electrons are attracted to this layer and complete the circuit back to battery B_1 through battery B_2. Notice that although it is usual to explain the action of the transistor in terms of electron flow, circuit symbols show directions of conventional currents which are in the opposite direction. The positive terminal of battery B_2 is at the top and conventional current flow is from positive to negative so that the arrow on the emitter points away from the device. Battery B_1 is connected so as to circulate current in an anticlockwise direction in its circuit.

Approximately 0.975 of the emitter current reaches the collector whilst the difference (0.025 of the emitter current) flows in the base. $I_C = \alpha I_E$ where α is defined as the current gain in the common base mode. (An alternative to the symbol α is h_{FB}).

As the emitter/base junction is forward biased it has very low input resistance and a very small change in battery B_1 voltage causes a large change in emitter current. The collector current therefore also suffers a large change. The collector/base junction is reverse biased and so has a high resistance. Now since battery B_2 voltage is considerably greater than that of B_1, although there is a small current loss through the device there is a large voltage gain and hence a power gain.

As an alternative to the npn transistor we may have a pnp device. The operation is similar except that all the current directions are reversed.

When comparing the BJT with the JFET we may relate emitter to source, collector to drain and to some extent base to gate although the BJT is current operated, there being a requirement for current in the base connection whilst the JFET is voltage controlled, the gate being held at a particular voltage but there is no current in the gate connection.

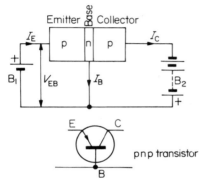

Figure 12.8

CIRCUIT CONFIGURATIONS

Referring first to the bipolar junction transistor, if an input signal is applied between the emitter and base terminals and the output is taken from the collector and base terminals so that the base connection is common to both input and output circuits, the transistor is said to be in common (or grounded) base configuration.

The equivalent configuration for the JFET is shown in *Figure 12.9b*. The gate connection is common to both input and output circuits. This circuit has very few practical applications however. *Figure 12.9c* and *d* show the common emitter and common source configurations for the two types of transistor respectively. The emitter or source is common to input and output circuits.

Let us now investigate the amplifying action of the common emitter circuit. (*Figure 12.9c*). In the common base configuration, with I_E = 1 mA, and the current gain $\alpha = 0.97$, 0.97 mA flow in the collector whilst 0.03 mA flow in the base.

Rearranging the circuit to common emitter configuration, for a current in the base of 0.03 mA, the emitter and collector currents will be closely as before, i.e. 1 mA and 0.97 mA respectively. In the con-

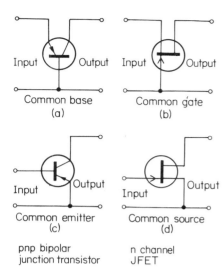

Common base
(a)

Common gate
(b)

Common emitter
(c)

Common source
(d)

pnp bipolar
junction transistor

n channel
JFET

Figure 12.9

figuration however, changing the input voltage changes the base current, and by varying this between zero and 0.03 mA the emitter current will be caused to vary between near zero and 1 mA with corresponding changes in collector current.

Note that I_B flows towards the base in an npn bipolar junction transistor and away from the base in the pnp type. When I_B is zero, I_C approaches zero. If the emitter/base voltage polarity is reversed, so attempting to reverse the direction of I_B the transistor switches off and there is no emitter or collector current. The BJT may therefore be used as a switch, being non-conducting or fully conducting according to its base/emitter polarity.

In the common emitter configuration

$$I_C = \beta I_B \quad \text{where } \beta = \frac{\alpha}{1 - \alpha}$$

$$\text{For } \alpha = 0.97, \quad \beta = \frac{0.97}{1 - 0.97} = 32.33$$

(An alternative to the symbol β is h_{FE}).

In this case the collector current is 32.33 times as great as the base current. The current gain through the transistor in the common emitter configuration is thus considerably greater than unity, a small base current controlling a much larger collector current. The voltage between the emitter and collector is much greater than that between the base and emitter (input voltage) so that with current and voltage gains, the power gain is considerable.

In the case of the JFET in the common source configuration, a small voltage input is made to control a current from a separate higher voltage source. Thus a relatively large power is controlled using virtually no input power since there is no gate current. (The resistance between the gate and channel is in the Megohm range.)

The greatest power gain using the bipolar junction transistor is obtained using the common emitter configuration whilst the highest voltage gain obtainable with the JFET occurs when using the equivalent common source configuration. These therefore are the usual connection modes for transistors in audio-frequency amplifiers.

The BJT may also be used in the common collector configuration while the equivalent for the JFET is the common drain circuit. In both cases the voltage gain is less than unity and these connections are used for special applications where a high input resistance and low output resistance are required.

SIMPLE SINGLE-STAGE TRANSISTOR AMPLIFIER CIRCUITS

There are many applications where a small alternating signal must be amplified into a larger one, the signal from a record player pick-up being made large enough to drive a load speaker for example.

This degree of amplification cannot be achieved using one transistor but rather by using several amplifier stages connected in series, each successive stage using the output from the previous stage as its input.

Considered here are two single-stage transistor amplifiers. The voltage changes which occur across the load resistor shown in *Figures 12.10* and *12.12* may be used to drive the next stage of a multi-stage amplifier or in these very simple cases, a pair of earphones for example.

Figure 12.10 shows a single npn transistor connected in the common emitter configuration and as we have already seen in this chapter, the

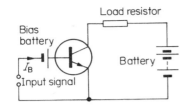

Figure 12.10

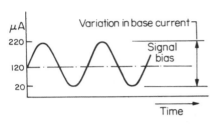

Figure 12.11

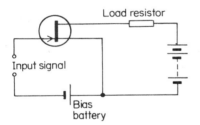

Figure 12.12

DECOUPLING CAPACITORS

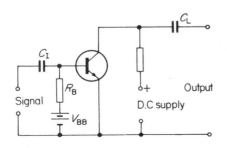

Figure 12.13

current flowing in the base of a BJT must always flow in the same direction. In the case of the npn transistor this direction is towards the base.

When an alternating signal is to be amplified it is necessary to provide a steady bias current in the base circuit, in the correct direction, of such a size that when the signal polarity reverses the base current continues to flow in the correct direction. In *Figure 12.10*, consider for example that the input signal e.m.f. has a magnitude sufficient to drive a sinusoidal current with a peak value of 100 µA. With no bias supply the transistor will only conduct when the signal current flows towards the base and during each reverse half cycle the transistor will switch off.

A bias voltage must be introduced of sufficient magnitude to drive a steady current in excess of 100 µA towards the base of the transistor. In *Figure 12.11* the bias current is shown with a value of 120 µA. When the signal voltage drives towards the base, the base current will rise to 120 + 100 = 220 µA and when the signal voltage reverses the base current will fall to 120 − 100 = 20 µA.

With no signal applied, the bias current in the base causes a much larger current to flow in the collector and this current flowing in the load resistor creates a steady potential difference across it. With the signal applied, the base current varies between 220 µA and 20 µA causing changes of considerably greater magnitude in the collector current with corresponding changes in voltage across the load resistor. Thus an alternating input signal superimposed on a bias gives rise to an alternating output current and load voltage also superimposed on steady values. The alternating power developed in the load resistor is considerably greater than that supplied to the input so that power amplification has been achieved.

Figure 12.12 shows a similar circuit employing an n-channel JFET. in this case since the voltage on the gate must always be negative with respect to the source, it is necessary to provide a negative bias voltage slightly greater in magnitude than the peak value of the signal voltage.

With bias voltage only, a steady drain current flows in the load resistor. The addition of an alternating voltage signal causes the drain current to alternate about its mean value as in the previous case. The alternating voltage developed across the load resistor is many times greater than that of the input signal so that considerable voltage amplification has been achieved. We cannot consider power amplification since in this case the power input is zero.

In the circuit shown in *Figure 12.10* it can be seen that the bias current provided by the bias battery must flow through the signal source. This may be avoided by using a decoupling capacitor. In *Figure 12.13* the required value of bias current is obtained by the adjustment of V_{BB} and R_B. The alternating voltage signal causes a displacement of charge on the capacitor C_I with a corresponding current flow in the emitter/base circuit and this is superimposed on the bias current.

In audio equipment only the changes in voltage and current output are useful since these cause a movement of a loudspeaker cone. The steady current in the output circuit, of whatever size, gives no audible output except perhaps a click as the circuit is energised and the loudspeaker cone moves into a position determined by the steady current.

Adding a further capacitor C_L allows only alternating current to flow in the load. The changes in voltage across the load resistor cause the capacitor charge and voltage to alternate and this charge displacement is reflected into the load, or where several stages of amplification are used, into the next stage input. At the same time the load circuit is isolated from the main d.c. supply.

A.C. INPUT RESISTANCE

This is the resistance offered to an alternating current by the input side of a device with specified conditions on the output side.

For the BJT in *Figure 12.10* in common emitter configuration, to measure the input resistance to alternating currents, the collector/emitter voltage V_{CE} is held constant while the base current is increased and decreased from the bias value over a range corresponding to the peak-to-peak value of the signal to be handled, for example from 220 μA to 20 μA as shown in *Figure 12.11*. This is achieved by adjusting the emitter/base voltage V_{EB}. A suitable circuit is shown in *Figure 12.14*.

$$\text{The a.c. resistance} = \frac{\text{Total change in } V_{EB}}{\text{Corresponding change in } I_B}$$

For a pure resistor the value is the same whatever the range used. The transistor in common with the semiconductor diode has a non-linear relationship between current and voltage so that the a.c. resistance has different values for different ranges and for different bias conditions. It also varies slightly as V_{CE} is changed.

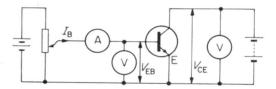

Figure 12.14

In common base configuration the a.c. input resistance has a different range of values and therefore particular note must be taken of values given with respect to the configurations and ranges for which they are valid.

In the case of the JFET in common source configuration, the a.c. input resistance is several megohms so that the gate current is virtually zero as already mentioned.

THE LOAD LINE FOR THE BIPOLAR JUNCTION TRANSISTOR

In *Figure 12.6* we saw the characteristic of the JFET in common source configuration. *Figure 12.15* shows the characteristics of a typical small-signal BJT in common emitter configuration and it can be seen that these are very similar in form to those for the JFET. Collector current I_C is plotted against collector/emitter voltage V_{CE} for a range of base currents between zero and 180 μA.

A graphical method to determine the output conditions from the input conditions for a particular value of load resistance involves the use of the 'load line' which is drawn on these characteristics. The load line intersects the characteristics enabling the collector current and the collector/emitter voltage to be obtained by inspection for any level of base current. This enables the current, voltage and power amplification to be determined provided that the input resistance is known. The use of the load line is best illustrated using an example.

Example (1).   Consider the simple amplifier stage shown in *Figure 12.16*. The load resistance = 1000 Ω and the main d.c. supply voltage = 10 V. The signal to be amplified is sinusoidal in form and has a peak value of 60 μA. The a.c. input resistance = 600 Ω.

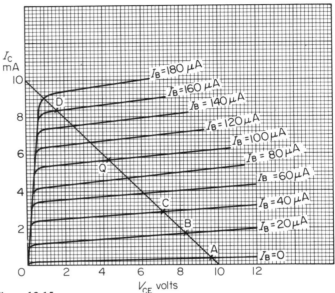

Figure 12.15

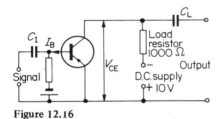

Figure 12.16

When $I_C = 0$ the voltage drop across the 1000 Ω load resistor = 0 Therefore, under these conditions $V_{CE} = 10$ V (the full supply voltage)

When $I_C = 10$ mA, the voltage drop across the load resistor = $10 \times 10^{-3} \times 1000 = 10$ V

$V_{CE} = 10 - 10 = 0$ volts.

For $I_C = 5$ mA, the voltage across the load resistor = 5 V.

Therefore $V_{CE} = 10 - 5 = 5$ V etc.

The load line is drawn across the characteristics as shown in *Figure 12.15*. It is a straight line passing through the points:

$V_{CE} = 10$ V when $I_C = 0$

$V_{CE} = 5$ V when $I_C = 5$ mA

$V_{CE} = 0$ V when $I_C = 10$ mA

In practice it is only necessary to fix the two end points, where V_{CE} is equal to the full supply voltage and zero respectively.

A bias current of at least 60 μA is required when the base current would vary between 60 + 60 = 120 μA and 60 − 60 = 0 μA. If the signal became slightly greater than 60 μA the transistor would shut off during part of the positive going half cycle. It is also necessary to consider the linearity of response of the transistor. For a change of 20 μA in the base current we require the same change in collector current at whatever level the change occurs, i.e. from 20 μA to 40 μA or from 100 μA to 120 μA etc. If this does not occur then the collector current waveform will not be an exact magnified copy of the base current and distortion will be present.

Measuring distances along the load line, the length AB in *Figure 12.15* as I_B changes from 0 to 20 μA is greater than the length BC for the next 20 μA change in I_B. The best, but still not perfect, spacing occurs between I_B = 40 μA and I_B = 160 μA. This means that the best bias value for this particular signal is 100 μA when the base current will vary between 160 μA and 40 μA. The point where the 100 μA characteristic and the load line intersect is marked Q and this is called the operating or quiescent point. At this point I_B = 100 μA, I_C = 5.6 mA and V_{CE} = 4.4 V.

As a check we see that V_{CE} = supply voltage $- I_C R_L$

$$= 10 - 5.6 \times 10^{-3} \times 1000 = 4.4 \text{ V}.$$

When the signal rises to its maximum value positive, the base current falls to 40 μA. The I_B = 40 μA characteristic and the load line intersect at C where V_{CE} = 7.1 V and I_C = 2.9 mA.

When the signal has a maximum negative value, the base current rises to 160 μA. The I_B = 160 μA characteristic and the load line intersect at D where V_{CE} = 1.7 V and I_C = 8.3 mA.

We are only concerned with changes in voltage and current which occur. The steady values are of no significance.

We are assuming sinusoidal waveforms throughout.

From Q the maximum signal swing is from D to C and this is the peak to peak value. The peak value is obtained by dividing by 2 and the r.m.s. value by a further division by $\sqrt{2}$.

The r.m.s. value of the signal = $\dfrac{120}{2\sqrt{2}}$ = 42.42 μA

V_{CE} changes from 7.1 V to 1.7 V due to this signal. Again this is peak to peak.

R.M.S. value of output voltage = $\dfrac{7.1 - 1.7}{2\sqrt{2}}$ = 1.909 V

Finally, I_C changes from 8.3 mA to 2.9 mA so that the r.m.s. value

$$= \frac{8.3 - 2.9}{2\sqrt{2}} = 1.909 \text{ mA}.$$

Only the a.c. components of current and voltage appear at the load due to the presence of the capacitor C_L

Signal voltage = signal current $\times$ input resistance

$$= 42.42 \times 10^{-6} \times 600$$

$$= 0.0254 \text{ V}.$$

Current gain = $\dfrac{\text{Load current}}{\text{Signal current}}$ = $\dfrac{1.909 \times 10^{-3}}{42.42 \times 10^{-6}}$ = 45.

$$\text{Voltage gain} = \frac{\text{Load voltage}}{\text{Signal voltage}} = \frac{1.909}{0.0254} = 75.$$

$$\text{Power gain} = \frac{\text{Power in load}}{\text{Power from signal}} = \frac{V \times I \,(\text{load})}{V \times I \,(\text{signal})}$$

$$= \frac{1.909 \times 1.909 \times 10^{-3}}{0.0254 \times 42.42 \times 10^{-6}}$$

$$= 3382.$$

Or power gain = voltage gain × current gain

$$= 75 \times 45 = 3375 \text{ (slight difference due to accumulated errors).}$$

Example (2).   Choose a suitable operating point and draw a load line on the characteristics given in *Figure 12.15* for the following conditions applied to the circuit shown in *Figure 12.16*.
The main d.c. supply is at 9.6 V.   Load resistance = 800 Ω.
The signal to be handled is of sinusoidal form and has a peak value of 40 μA.
The a.c. input resistance = 550 Ω.
Using the load line determine: (a) the current gain, (b) the voltage gain and (c) the power gain.

INCORRECT CHOICE OF OPERATING POINT

Suppose instead of choosing point Q for the operating point we had chosen point C in *Figure 12.15* when I_C = 2.9 mA. Considering the same signal as in Example 1, the variation in the value of I_C is shown in *Figure 12.17*. As soon as the signal reaches a positive-going value of 40 μA the base current and I_C fall to zero and remain at this value until the signal amplitude falls below 40 μA once again. The waveform of I_C and hence of the alternating voltage across the load resistor is not sinusoidal and it is this type of distortion which gives small radio sets the characteristic 'out-of-a-box' sound.

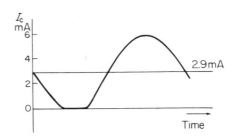

Figure 12.17

Whatever operating point is chosen, if the signal becomes too large for the device cut-off occurs at I_B = 0 and in addition, the top half of the wave also becomes distorted. The peak-to-peak range of the signal must lie well within the range of base currents which the transistor can handle.

THE LOAD LINE FOR THE JFET

Figure 12.18 shows the common source JFET circuit and *Figure 12.19* its characteristics. The procedure is generally as for the BJT.

Example (3). Draw the load line for the JFET in the circuit shown in *Figure 12.18* on the characteristics shown in *Figure 12.19*. The signal to be handled is sinusoidal in form and has a peak value of 0.4 V. The d.c. supply voltage = 9.6 V. The load resistance = 800 Ω. Determine the value of voltage gain and the load power.

When $I_D = 0$, $V_{DS} = 9.6$ V (the full d.c. voltage.)

When $I_D = \dfrac{9.6}{800} = 0.012$ A, $V_{DS} = 0$

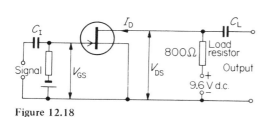

Figure 12.18

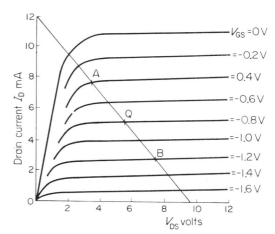

Figure 12.19

The load line is drawn through the points: $V_{DS} = 0$, $I_D = 12$ mA
$V_{DS} = 9.6$ V, $I_D = 0$.

An operating point is chosen on the load line in such a position that V_{GS} remains negative at all times and at a point in the characteristics where the greatest linearity of response to changes in V_{GS} is to be found.

A suitable point for Q is where the load line crosses the $V_{GS} = -0.8$ V characteristic.

As the signal swings ± 0.4 V the limits of V_{GS} are -0.4 V and -1.2 V.

At $V_{GS} = -0.4$ V, $V_{DS} = 3.5$ V, $I_D = 7.6$ mA (point A)

At $V_{GS} = -1.2$ V, $V_{DS} = 7.5$ V, $I_D = 2.6$ mA (point B)

The r.m.s. value of the signal = $\dfrac{0.4}{\sqrt{2}} = 0.283$ V

The r.m.s. value of the output voltage $= \dfrac{7.5 - 3.5}{2\sqrt{2}} = 1.41$ V

The r.m.s. value of the output current $= \dfrac{7.6 - 2.6}{2\sqrt{2}} = 1.77$ mA.

Voltage gain $= \dfrac{1.41}{0.283} = 5$

Load power $= 1.41 \times 1.77 \times 10^{-3}$

$\qquad\qquad = 2.5$ mW.

As already stated in this chapter, the power input is virtually zero so that power gain is not applicable.

Example (4). Draw a load line on the characteristics given in *Figure 12.19* for the following conditions applied to the circuit shown in *Figure 12.18.*

Main d.c. supply voltage = 12 V.
Load resistance = 1200 Ω.
The input signal is sinusoidal in form and has a peak value of 0.2 V.
Select a suitable Q point and determine (a) the value of voltage gain, (b) the r.m.s. value of the drain current, (c) the power developed in the load resistor.

Note that all the comments concerning the choice of operating point made after the load line analysis for the BJT apply to the JFET, the bias and signal being voltages instead of currents.

HIGH AND LOW FREQUENCY EFFECTS

The voltages applied to transistors create barrier layers at reverse biased junctions. Each of these barrier layers is a good insulator and a capacitor is formed with the depletion layer as the dielectric and the adjacent conducting layers as the plates. The width of the depletion layer and hence the value of capacitance of the capacitor formed depends on the applied voltage as seen particularly in the case of the JFET. The effect of the capacitance is to allow the a.c. components of current to take paths other than those intended so shunting part of the input signal and output current.

As the frequency increases, the value of the capacitive reactance falls so that the shunted currents increase and the output falls. At a particular frequency dependent on the design of the device a frequency is reached at which no amplification takes place.

Where decoupling capacitors are used there will be a low-frequency cut-off since at very low frequencies the capacitive reactance of these capacitors will be large.

EXPERIMENTAL WORK

Using the characteristics and load line for a particular transistor, as shown in the earlier part of the chapter, an operating point may be chosen. The choice of operating point may be justified using the circuit shown in *Figure 12.20.*

This is a test circuit for a BJT. The circuit for the JFET is identical except for d.c. supply polarities (see *Figure 12.18*). The ammeter to measure I_B will be superfluous.

It is only possible to give general guidance on values, the manufacturers' data for the particular transistor will have to be consulted before connecting supplies. This will give maximum voltages and currents for the device.

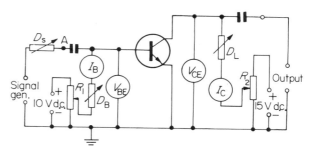

Figure 12.20

With the signal generator switched off and D_L set to zero, R_2 is used to set V_{CE} to the required value. With D_B set to several thousand ohms, R_1 is adjusted to give the steady base current as indicated by the Q-point on the characteristics. D_L is adjusted to the required value for the load. The steady state conditions have now been created.

The signal generator is switched on and a sine wave of the desired frequency applied to the input. A CRO is used to view the input waveform between point A and the bottom line (earth). A second CRO trace is used to view the output voltage. The input signal voltage may be increased in magnitude until the output voltage waveform becomes distinctly non-sinusoidal.

The effect of bias conditions may be examined by altering the setting on R_1. If the bias is too small the bottom of the output voltage waveform will be clipped (see *Figure 12.17*), while too large a bias will result in the top being clipped. Distortion is minimised by careful adjustment of the bias and input signal amplitude.

Voltage gain is determined by comparing the amplitudes of input and output voltage waveforms.

PROBLEMS FOR SECTION 12

(5) A bipolar junction transistor in common base configuration is carrying a steady base current of 0.05 mA. $\alpha = 0.99$. Determine the values of (a) the emitter current, (b) the collector current.

(6) A bipolar junction transistor with $\beta = 60$ is carrying a steady collector current of 5 mA. Calculate the values of (a) the emitter current and (b) the base current.

(7) Why is the BJT most often used in the common emitter configuration?

(8) Why is there an upper and lower frequency limitation on the operation of a transistor amplifier?

(9) Why is it necessary to provide a bias voltage or current when using a transistor to amplify alternating signals?

(10) Describe how a BJT may be used as an on/off switch.

(11) What is the function of a decoupling capacitor?

(12) When testing a simple amplifier stage employing a BJT in common emitter configuration as shown in *Figure 12.16*, the following readings were obtained:

Input resistance = 850 Ω.

Bias current = 25 μA.

Peak base current = 40 μA. Minimum base current = 10 μA.

Minimum value of V_{CE} = 6.8 V.   Maximum value of V_{CE} = 8 V.

Minimum value of I_C = 1.3 mA.   Maximum value of I_C = 4 mA.

Determine: (a) the current gain, (b) the voltage gain and (c) the power gain of the stage.

(13) When testing a BJT in common emitter configuration, with V_{CE} held constant, a change in V_{BE} from 0.06 V to 0.17 V caused the base current to change from 100 μA to 220 μA. What is the value of the a.c. input resistance of the device under these conditions. Why might a different value be obtained for different changes in V_{BE}?

(14) What is the effect on the output current and voltage waveforms of (a) insufficient bias and (b) too large an input signal?

(15) The data in the table refers to a BJT in the common emitter configuration.

Collector/emitter voltage V_{CE} volts	Collector current (mA)		
	Base current 20 μA	Base current 70 μA	Base current 120 μA
1	0.8	3.3	5.8
3	1.2	3.8	6.4
5	1.6	4.4	7.1
7	2	4.9	7.7
9	2.4	5.4	8.35

Draw a load line on these characteristics for a d.c. supply voltage of 8 V and a load resistance of 800 Ω.

The input resistance = 1000 Ω.

Determine for a sinusoidal signal of 50 μA peak value:

(a) the r.m.s. value of the input current,

(b) the r.m.s. value of the load current,

(c) the r.m.s. value of the load voltage,

(d) the current gain,

(e) the voltage gain and

(f) the power gain.

13 High power electronics: the silicon controlled rectifier (SCR)

Aims: At the end of this section you should be able to:
Describe the structure, and draw the circuit symbol of the SCR
Draw a set of static characteristics.
Explain the operation of the SCR
Describe simple phase shifting gate control circuits.
Draw waveforms for gate current, anode/cathode voltage and load voltage at different trigger angles.
Describe the use of the SCR as an invertor and convertor.

THE DEVELOPMENT OF THE SCR

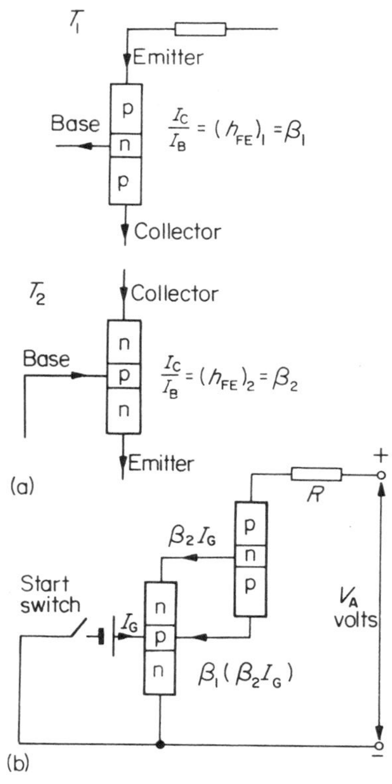

$$\frac{I_C}{I_B} = (h_{FE})_1 = \beta_1$$

$$\frac{I_C}{I_B} = (h_{FE})_2 = \beta_2$$

(a)

(b)

Figure 13.1

Consider the two transistors shown in *Figure 13.1*.

T_1 is a pnp bipolar junction transistor whilst T_2 is an npn type. They have current gains in the common emitter configuration of β_1 and β_2 respectively. As shown in *Figure 13.1a*, T_2 needs a current inwards at its base to become conducting and T_1 a current outwards from its base.

If a current I_G is injected into the base of T_2 whilst its collector is at a positive potential with respect to the emitter, the collector current will rise to $\beta_2 I_G$. The collector is connected to the base of T_1 so that its base current becomes $\beta_2 I_G$. Again provided that the polarity of the supply is correct, the collector current of T_1 will rise to $\beta_1(\beta_2 I_G)$. This will be many hundreds of times greater than I_G. Looking at *Figure 13.1b* we see that this current is fed back to the base of T_2 where it will cause a further increase in the collector current and hence in the base current of T_1. In a very short time the device becomes a short circuit, and since the forward volt drop is very small the current flowing is limited only by the circuit resistance (R in *Figure 13.1b*).

If the polarity of the applied voltage V_A is reversed, no current flows so that the device can function as a rectifier or unidirectional switch. Conduction may be caused to commence at any instant during the positive going half-cycle of an a.c. voltage wave by the provision of an input to the base layer of the npn transistor.

Once the gate pulse has been supplied and the through current is flowing, no variation in the gate current or its polarity will cause the device to shut off. This can only be achieved by reducing the applied voltage V_A to zero, or driving it negative, for a short period. When acting as a rectifier, such a condition occurs at the end of each positive going half cycle. To cause conduction to commence gate current must be provided at a suitable instant during each positive going half cycle.

THE PRACTICAL THYRISTOR

The thyristor so far considered is perfectly satisfactory from an operational point of view but is larger than necessary. The device is normally formed on one chip of semiconducting material. From an inspection of

Figure 13.1 we see that the base layer of transistor T_1 and the collector layer of transistor T_2 are joined together and they are both n-type material. These two layers could in fact be made common. Similarly the T_1 collector p-type material could be made common with the T_2 base p-type layer.

Figure 13.2a shows a four layer arrangement which performs in the same manner as the two transistors.

Figure 13.2b and *c* show the arrangement of the layers on the chip and the chip in its capsule.

Currents of several hundred amperes can be handled and with a forward voltage drop of 1 or 2 V, the power loss may be several hundred watts in the larger sizes when a heat sink of considerable proportions is required.

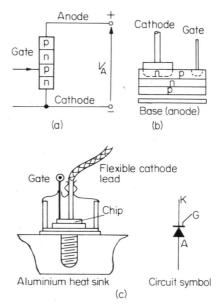

Figure 13.2

THYRISTOR CHARACTERISTICS

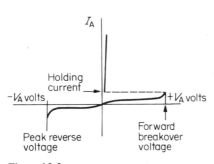

Figure 13.3

The thyristor will break down and become conducting in either direction if sufficient voltage is applied. *Figure 13.3* shows the forward and reverse characteristics with no gate current. Steadily increasing the anode voltage in the forward direction causes little current to flow until the forward breakover voltage is reached when the anode current suddenly increases and the voltage drop across the device falls to a very low value. The current flowing immediately after the breakdown is called the holding current. This is a value of anode current below which the thyristor may turn off. A circuit current slightly in excess of this value must be maintained once conduction has started in order to ensure that the device does not return to the non-conducting or off state.

If the anode voltage polarity is reversed the thyristor becomes conducting when the peak reverse or peak inverse voltage is reached. It is common practice to increase the capability of the SCR to resist reverse breakdown by running it in series with a silicon diode which has the effect of restricting the flow of reverse current at any voltage as so reducing the power loss and the heat produced.

Figure 13.4 shows the effect on the forward characteristic of increasing the gate current. If a gate current in excess of 50 mA is used for this particular device, the forward voltage drop is extremely small and the device becomes conducting almost immediately. With smaller gate currents the anode current increases gradually as the voltage V_A is increased until at a particular value the device becomes conducting as in *Figure 13.3*. During this prolonged switching-on period the power loss is high and the device rapidly overheats.

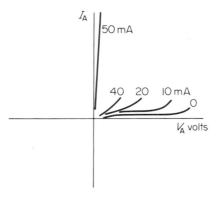

Figure 13.4 Forward characteristics for gate currents between zero and 50 mA

The desired shape of a current pulse to trigger a thyristor is one which has a rapid rate of rise to the required value to cause conduction. The duration of the pulse depends on the type of load being controlled. The thyristor remains conducting after a very short pulse if the current flowing in the anode circuit is by then in excess of the holding value. This value is quoted by the manufacturer in his catalogue. With a resistive load the thyristor current rises very rapidly and a triggering pulse of a few microseconds duration is sufficient. Where the load is inductive and the time constant of the circuit is long, the gate pulse may have to last much longer to allow time for the circuit current to reach the holding value.

SIMPLE CONTROL CIRCUITS FOR SINGLE-PHASE OPERATION

Figure 13.5 shows a very simple circuit for the control of a single thyristor. It is not possible to predict exactly the instant at which conduction will commence when using this method.

The potential at point G in *Figure 13.5* will be a function of the charge received by capacitor C through resistor R. During a positive-going half-cycle, current flows through R charging C and at a particular value of voltage; dependent on diode D_1 and the thyristor characteristics; sufficient current will flow through D_1 to trigger the thyristor. The gate current is limited by R once the initial discharge of C has occurred.

During a negative half-cycle the capacitor C charges with reverse polarity, R being bypassed by diode D_2. Diode D_1 prevents current flowing from the gate during this period to prevent damage to the thyristor.

At the commencement of the next positive half-cycle the top plate of the capacitor starts strongly negative and has to become positive before the thyristor can become conducting.

A method which produces steep-fronted pulses and so triggers the thyristor at a precise instant is shown in *Figure 13.6*. A peaking trans-

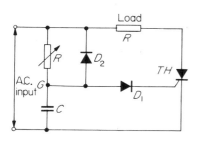

Figure 13.5

former is used. This has a ferrite core which saturates suddenly at a fairly low level of flux density.

Considering a sinusoidal input voltage to the transformer: As the magnetising current increases and the core flux increases, a voltage is induced in the secondary. The core quickly saturates and for much of the half-cycle there is no further change in flux. Voltages are induced only by changing fluxes so that the secondary voltage falls to zero. Voltage will be induced in the secondary at the end of the positive half-cycle and the beginning of the negative half-cycle as the flux is reduced and then reversed followed by a further period of zero voltage. A spiky voltage waveform is produced as shown in *Figure 13.6*.

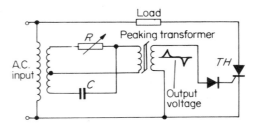

Figure 13.6

The phase of the input voltage to the peaking transformer is varied through very nearly $180°$ by varying R between zero and a very large value. Thus a triggering pulse may be provided at any instant during the positive-going half-cycle. (A similar pulse occurs as the voltage goes negative but this is only useful for full-wave operation using two thyristors.)

Figure 13.7 shows a phase-shifting transformer providing the required phase shift. A phase-shifting ransformer is very similar to an induction motor except that the rotor is not allowed to rotate freely. The stator carries two windings which are displaced by $90°$. One winding is connected directly to the supply while the other is fed through a capacitor. In this manner the two winding currents are made to be $90°$ out of phase electrically.

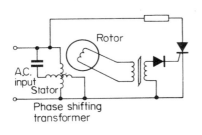

Figure 13.7

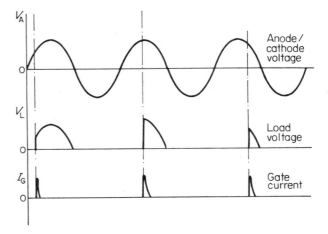

Figure 13.8

The magnetic field created by the stator links with a single coil on the rotor. By turning this through 360°, the phase of the induced voltage may be varied by this amount, the rotor coil first linking mainly with the flux from one stator coil and then the other and then with each of them in turn with reversed direction and hence polarity as the rotation continues. The output from the rotor is fed to a peaking transformer so that pulses of the correct shape are obtained as required throughout 360°.

Figure 13.8 shows the effect of providing the gate pulse at different instants during the positive half-cycle of a sinusoidal voltage wave applied between the anode and cathode of a thyristor connected to feed a resistive load. These waveforms may be viewed using a CRO, connecting the input leads to the relevant terminals in the circuit. Care must be taken however to see that by so doing, parts of the circuit are not inadvertently earthed. If one of the CRO terminals is earthed, connecting it across the load terminals for example, to view the load voltage waveform, will cause either the main input or the anode of the thyristor to become earthed with possibly disastrous results. This may occur at other points in the circuit and danger is avoided by using an earth free oscilloscope or an isolating transformer either to feed the main equipment or to isolate the CRO input from the main circuit.

FULL-WAVE OPERATION (A.C. IN LOAD)

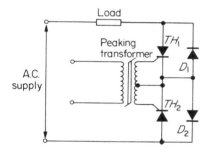

Figure 13.9

Load current is obtained during both positive and negative half-cycles of the supply voltage by using two thyristors. In *Figure 13.9* the input to the peaking transformer is provided from an *RC* dividing network as in *Figure 13.5* or a phase-shifting transformer as in *Figure 13.6*. A positive-going pulse causes current to flow towards the gate of TH_1 which becomes conducting and current flows in the load from the supply, through TH_1 and D_2. At the end of the positive half-cycle conduction ceases. A pulse in the opposite direction will cause current to flow towards the gate of TH_2 and load current flows from the supply through TH_2 and D_1 during the negative half-cycle.

Figure 13.10 shows the full-wave output. It may be monitored using the CRO as in the half-wave case.

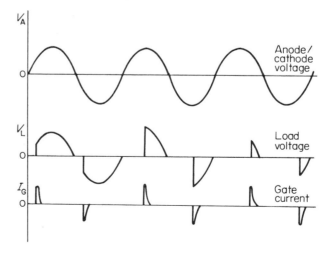

Figure 13.10

FULL-WAVE CONTROLLED RECTIFICATION

Where the load is to be provided with a controlled level of d.c. the circuit shown in *Figure 13.9* is modified to form the bridge circuit which gives full-wave controlled rectification. The load voltage waveforms are then as shown in *Figure 13.8* with the addition of a conducting period during each negative half-cycle.

Alternatively the biphase connection may be used (*Figures 11.3* and *11.4*).

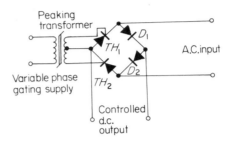

Figure 13.11

OTHER APPLICATIONS OF THE SCR

The chopper   *Figure 13.12* shows a simple circuit of a chopper which enables the power being supplied to a load from a d.c. supply to be varied without the use of series resistors. This is particularly useful in battery electric vehicles where distance travelled between charges is very important. Acceleration and speed control are achieved with virtually no power loss and if regenerative braking is used, so charging the batteries as the vehicle runs downhill or stops, such a system of control can more than double the distance travelled between charges when compared with rheostatic systems.

Consider both thyristors to be off. TH_1 receives a trigger pulse and current flows in the load, TH_1 and one half of the centre tapped inductor. The sudden increase in current gives rise to induced voltages in both halves of the inductor since all of the turns are linked with the changing flux. The direction of the induced e.m.f. together with the resulting charge on the capacitor C are shown in *Figure 13.12*.

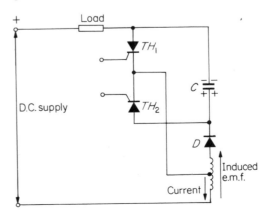

Figure 13.12

If now TH_2 is triggered, the cathode of TH_1 is connected directly to the positively charged plate of the capacitor and is thereby switched off. (The requirement for switching off a thyristor is to remove or reverse the anode/cathode potential.)

Load current ceases and can be re-established by triggering TH_1 again. The triggering pulses are generated by an oscillator and power in the load is controlled by varying the 'on' time, i.e. the interval between triggering TH_1 and TH_2, and by altering the frequency at which switching is carried out.

The invertor
The invertor is used to produce alternating voltages and currents using a d.c. source of power. This is useful for example for operating fluorescent lighting and other a.c. equipment in vehicles and caravans using a battery source or as part of a convertor which produces a.c. of a particular frequency while being supplied with a.c. of another frequency.

The mode of operation is very similar to that of the chopper. In *Figure 13.13*, starting with both thyristors off; TH_1 is triggered and current flows from the supply through half of the transformer primary winding and back to the supply negative. The increasing flux produced

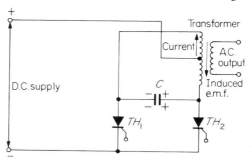

Figure 13.13

by the rising current induces voltages in both the primary and secondary windings of the transformer. The direction of this e.m.f. and the resulting charge on capacitor C is shown in *Figure 13.13*. Triggering TH_2 effectively connects the positive plate of the capacitor to the cathode of TH_1 which switches off. Current now flows from the supply through the second half of the transformer winding so that the polarity of the induced e.m.f. and the charge on the capacitor are reversed. Triggering TH_1 and TH_2 in a regular sequence, again using an oscillator, causes a periodically varying voltage to be induced in the secondary winding of the transformer. A.C. has been produced using a d.c. source.

The convertor
A.C. of one frequency is rectified and the resulting d.c. fed to an invertor which operates at the required output frequency. Thus two a.c. systems of different frequencies may be interconnected or a.c. of any required frequency may be produced using an a.c. source of another frequency.

PROBLEMS FOR SECTION 13

(1) Explain the operation of the SCR in terms of the two transistor model.

(2) Define the term 'Holding-current' and explain what can happen if at any time the anode current in an SCR falls below this value.

(3) What advantages are there to using an SCR as a switch as compared with a mechanical contactor?

(4) Draw a circuit diagram showing a simple phase-shifting circuit suitable for the control of an SCR.

(5) What dangers are there in using an earthed CRO when viewing the voltage waveforms in an SCR circuit? How are these dangers overcome?

(6) Draw circuits showing thyristors controlling power from an a.c. supply (a) into an a.c. load circuit, (b) into a d.c. load circuit.

(7) What advantages are there to using SCRs to control d.c. power as compared with rheostatic methods?

14 Thermionic devices

Aims: At the end of this section you should be able to:
Describe thermionic emission from cathode surfaces.
Draw, and describe the action of, vacuum diode, triode and pentode valves.
Describe gaseous ionisation.
Describe the action of the gas-filled diode and triode (thyratron).
Use the load line together with triode characteristics to determine the voltage gain of a triode amplifier.
Describe the construction of the cathode-ray tube and show how the electron beam is controlled by electric or magnetic fields.

THERMIONIC EMISSION

Atoms consist of a nucleus and orbiting electrons and in electrical conductors the electrons furthest from the nucleus are able to move in a random manner from atom to atom. The mobility of these electrons increases as the conductor is heated and at a particular temperature dependent on the material, some of them will have attained sufficient energy to escape completely from the restraining forces exerted by the nuclei.

If the heating takes place in a near perfect vacuum the liberated electrons can move away from the heated surface but with air present there are collisions between the electrons and the much larger molecules of air which use up the extra energy and the electrons return to the surface.

To be useful in electronic valves, electrons must be liberated in large numbers at a temperature below that at which the material melts. Pure tungsten metal will emit electrons into a vacuum when heated to about 2500°C. This uses a considerable amount of power and the emission achieved represents only about 3 mA per watt of heater power. At around 2000°C the emission is virtually zero. Adding a small amount of thorium oxide to the tungsten increases the emission considerably. Oxides of calcium, strontium and barium give excellent emission at temperatures around 1000°C.

THE DIODE VALVE

Figure 14.1 shows a vacuum diode. It consists of a cathode, which is heated to incadescence, surrounded by a metal cylinder called the anode; in a glass envelope exhausted to near-perfect vacuum. The cathode emits electrons which form a space charge around it. Since electrons are negatively charged, this cloud of electrons is very strongly negative, and once established repels further electrons trying to leave the cathode surface, preventing their doing so since like charges repel one another.

In a circuit as shown in *Figure 14.2*, the anode being positive with respect to the cathode, it attracts electrons from the space charge making it weaker. More electrons are emitted from the cathode to replace them and a current is established round the circuit.

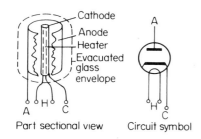

Part sectional view   Circuit symbol

Figure 14.1

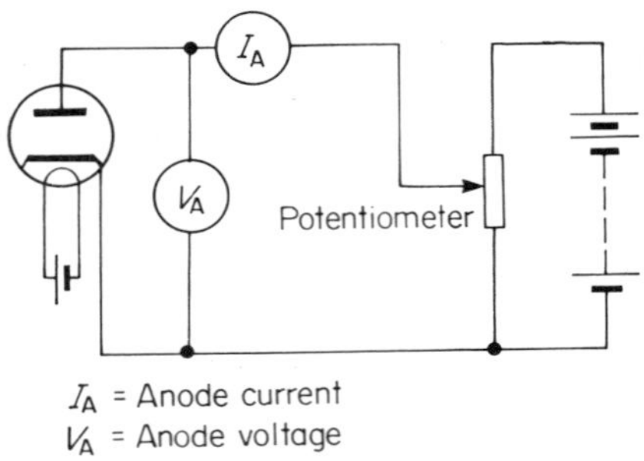

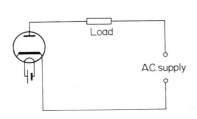

I_A = Anode current
V_A = Anode voltage

Figure 14.2

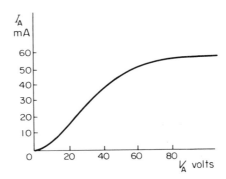

Figure 14.3

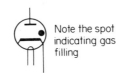

Figure 14.4

As the voltage at the anode is increased from zero to 20 V the anode current increases very nearly proportionally as shown in *Figure 14.3*. A further 20 V increase in anode voltage produces a corresponding change in anode current. However, for this particular valve, increasing the anode voltage above about 60 V produces very little change in anode current. The reason for this is that the whole of the space charge has been attracted across to the strongly positive anode and it is now possible to attract electrons only at the same rate as they are emitted from the cathode, and this is a function of the cathode temperature. Hence for a particular temperature there is a saturation current. The saturation current can be increased for a given cathode only by increasing its temperature and there is a limit to this since it will ultimately melt.

Connecting the diode as shown in *Figure 14.4*, using an a.c. supply, current flows during the half-cycle when the anode is positive but not when it is negative since then it repels electrons. This is then a half-wave rectifier.

THE GAS-FILLED DIODE

Note the spot indicating gas filling

Figure 14.5

In the diode valve, with the anode strongly positive, the electrons move through the cathode/anode space at high velocity and are hurled against the anode probably disappearing within it. The kinetic energy of the electron heats up the anode.

If a small quantity of gas is introduced into the space, the high velocity electrons collide with the gas atoms during their travels. The electrons, being much lighter than the atoms of gas, are deflected from their original paths whilst the atoms are virtually unmoved (consider the effect of a small pebble being thrown at a rock).

If the impact is sufficiently violent the result of a collision is to liberate an electron from the surface of the gas atom which, since it is now short of one electron, exhibits positive charge and is called a *positive ion*. The liberated electron is accelerated towards the anode and may itself be involved in collisions liberating further electrons. (The ions formed move towards the cathode and there receive electrons turning them into complete gas atoms again.)

The gas in the envelope is said to be ionised and ionisation produces light. Different gases and pressures produce light of different colours. In the gas-filled diode valve, argon or xenon gas or a little mercury vapour are used.

Once ionisation has taken place, the internal resistance of the valve falls to a very low value and a typical characteristic is shown in *Figure 14.6*.

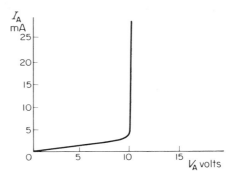

Figure 14.6

For values of V_A up to about 10 V, I_A increases in the same manner as in the vacuum valve. However for a value of V_A marginally above 10 V the current increases to a value determined mainly by the resistance of the external circuit. A glow appears inside the diode, the brilliance of which is a function of the current flowing. It is essential to have external resistance in the circuit to limit the current.

When used as a rectifier, virtually all the supply voltage is available at the load and much larger currents can be handled than with the vacuum valve.

COLD-CATHODE VALVES

The cathode/anode space can be ionised by the application of a sufficiently high voltage to provide the energy to liberate electrons from the cathode and so start the process. A cold-cathode valve used as a voltage stabiliser is shown in *Figure 14.7*. In the region of 150 V the valve suddenly becomes conducting and acts in a similar manner to the zener diode (see Chapter 11).

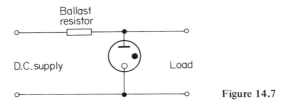

Figure 14.7

Another cold-cathode device is the mercury discharge lamp as shown in *Figure 14.8*.

An initial discharge takes place between the auxiliary starting electrode and one cathode ionising a small quantity of the mercury vapour present in the tube. The ions and electrons so formed rapidly move down the tube and a discharge takes place between the two cathodes. The subsequent collisions with the cathodes of electrons and ions

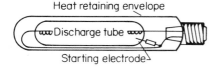

Figure 14.8

keep them hot enough to maintain the discharge, if the tube is mounted in a heat retaining envelope. Since the internal resistance is very low when running, a series-connected current-limiting choke has to be employed.

THE VACUUM TRIODE VALVE

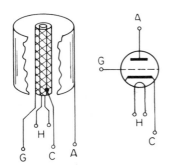

Figure 14.9

Figure 14.9 shows a part sectional view of a vacuum triode valve and its circuit symbol. The triode has three electrodes, a cathode and anode as in the diode valve, together with a third called the control grid which consists of a fine wire mesh or helical winding through which electrons can pass. The control grid is very close to the cathode and is supplied at a potential which is negative with respect to the cathode. Due to the close spacing, the grid/cathode voltage, V_G has much more effect on the anode current than has the anode/cathode voltage, V_A.

For a particular value of V_A, with V_G set to zero, the valve acts as a diode and in *Figure 14.10* we see that when V_A = 60 V and V_G = 0, I_A = 50 mA. Making the control grid negative with respect to the cathode neutralises in part the effect of the positive anode and I_A falls. The control grid acts as a variable space charge so limiting the cathode emission. Increasing the magnitude of V_G from zero to −4 V causes I_A to fall to zero when the effect of +60 V on the anode is exactly neutralised by the effect of −4 V on the grid. The value of V_G which causes

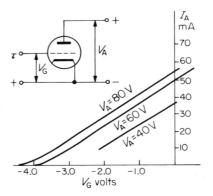

Figure 14.10

I_A to fall to zero is known as the cut-off value (compare this with pinch-off value for the JFET in Chapter 12).

Again with V_G set to zero but with V_A increased to 80 V, I_A = 54 mA. Such an increase in anode current would be expected from a consideration of the diode characteristics in *Figure 14.3*. To cause the anode current to fall to zero with this greater anode voltage applied requires the magnitude of the cut-off voltage to be increased to approximately 4.5 V.

Figure 14.10 shows what are called the 'mutual characteristics'. *Mutual* means the effect of change in one circuit (the grid circuit) on the current in the other circuit (the anode circuit). These mutual characteristics for a triode valve are very similar to those for the JFET (*Figure 12.5*). Both devices use a voltage (grid or gate) to control a current (anode or drain).

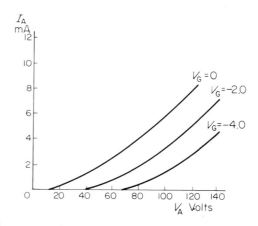

Figure 14.11

With the triode valve, keeping V_G constant and varying V_A over the working range whilst measuring I_A enables a set of anode characteristics to be plotted as shown in *Figure 14.11*.

The triode voltage amplifier

[Biasing and the use of the decoupling capacitor are dealt with fully in Chapter 12 and these sections should be read and understood before proceeding here.]

Figure 14.12 shows a triode valve connected as a single-stage amplifier. The control grid has two voltages, connected in series, applied to it. One of these is a bias voltage provided by a battery, and the other is the alternating signal to be amplified. The grid bias battery is connected with the polarity shown and has an e.m.f. of magnitude such that with the signal applied the grid of the valve can never become positive. Should this occur some of the electrons bound for the anode are attracted to it and have to flow back to the cathode through the signal circuit. This causes distortion of the input. In addition, the grid is not designed to carry current and it could overheat.

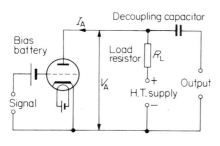

Figure 14.12

The alternating grid potential gives rise to changes in I_A and these produce voltage changes across the load resistor. The alternating output is isolated from the high-voltage supply by using a decoupling capacitor.

Although the changes in I_A are generally only in the milliampere range, the load resistor has a value of several thousand ohms so that the alternating component of voltage across it is much larger than the signal voltage and voltage amplification has been attained.

THE LOAD LINE

The construction and use of the load line for the triode amplifier follow closely the procedures described in Chapter 12 for transistors and it is recommended that the relevant parts of that chapter are read in conjunction with this section.

Example (1). A triode valve with anode characteristics as given in *Figure 14.13* is to be used to amplify a sinusoidal signal with peak value 2 V. The value of the load resistor is 20 000 Ω, and the h.t. supply is at 300 V.

Draw a load line and using a suitable operating point, determine the value of voltage amplification obtained.

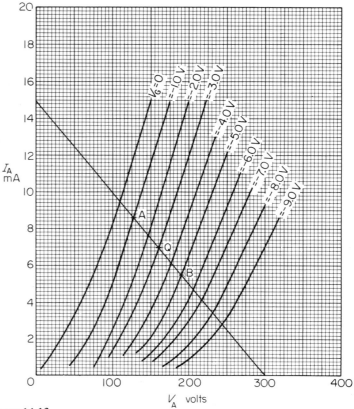

Figure 14.13

When $I_A = 0$, $V_A = 300$ V (The full supply voltage)

When $I_A = \dfrac{300}{20\,000} = 15$ mA, $V_A = 0$.

The load line is drawn between the points: $V_A = 0, I_A = 15$ mA

and $V_A = 300$ V, $I_A = 0$.

Since the signal has a peak value of 2 V, a suitable bias value would be −3 V. The operating point, Q is shown on the $V_G = -3$ V characteristic in *Figure 14.13*.

The grid voltage swings between −3 −2 = −5 V and −3 +2 = −1 V. The corresponding values of V_A are 190 V (point B) and 128 V (point A).

The r.m.s. value of the signal $= \dfrac{2}{\sqrt{2}} = 1.414$ V.

The r.m.s. value of the output voltage $= \dfrac{190 - 128}{2\sqrt{2}} = 21.92$ V.

Voltage amplification $= \dfrac{21.92}{1.414} = 15.5$.

Example (2).   Using the load line in *Figure 14.13* and a signal with 2 V peak value, determine the value of voltage amplification using (a) −2 V bias, (b) −7 V bias.

Example (3).   Using the characteristics in *Figure 14.13*, determine the voltage amplification using a load resistor of 12 500 Ω and an h.t. supply at 250 V when the signal has a peak value of 3 V.

THE TETRODE AND PENTODE VALVE DEVELOPMENT

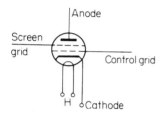

Figure 14.14

In a triode valve amplifier, as the grid potential becomes less negative the anode current increases causing an increased voltage drop across the load resistor. The anode voltage therefore falls. Hence the anode current is less than it would be if the anode voltage were maintained constant.

The tetrode valve was developed to overcome this problem. It has an additional screen grid which is maintained at constant positive potential with respect to the cathode so that whatever changes occur in the anode voltage, electrons coming through the control grid come under the influence of a constant accelerating voltage.

When the anode voltage is zero all the electrons emitted from the cathode and up on the screen grid since there is no reason for them to travel any further. As the anode potential is increased electrons reach the anode and I_A is a function of V_A as in the triode. This is region A in *Figure 14.15*.

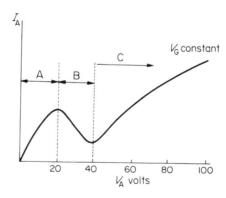

Figure 14.15

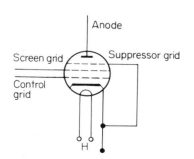

Figure 14.16

Further increases in anode potential cause the electrons to reach sufficient velocity that as they strike the anode secondary electrons are emitted from its surface, and since these are travelling at a slower speed than those originally striking the anode, they are attracted back to the screen grid since this is at a higher potential than the anode at this point in the characteristic. This is region B in *Figure 14.15*.

As the anode voltage is increased still further it becomes large enough to attract back the secondary electrons and the total current increases as in region C in *Figure 14.15*.

This characteristic is useful in certain applications but not in amplifiers, since over the range 20–40 V on the anode a falling anode current is obtained for increasing voltage.

The distortion of the tetrode characteristic can be overcome by the addition of a further electrode when the valve becomes a pentode. The addition to the valve is the suppressor grid and it is connected directly to the cathode.

The slow moving electrons which may be emitted from the anode due to impact will always be attracted back to the anode since they will be repelled in that direction by the suppressor grid which under all conditions is negative with respect to the anode. It might be argued that the suppressor grid would prevent the original electrons emitted from the cathode reaching the anode. However the screen grid voltage is strongly positive with respect to the cathode and this causes the electrons to attain sufficient velocity that they pass through it and the suppressor grid without appreciably slowing down.

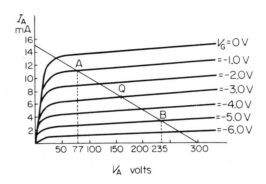

Figure 14.17 Pentode anode characteristics

The anode characteristics of the pentode valve are shown in *Figure 14.17*.

A load line for an h.t. supply of 300 V and a load resistor of 20 000 Ω is shown and the voltage amplification under these conditions $(235-77)/4 = 40$ for a signal with a peak value of 2 V as compared with 15.5 in Example 1.

It can be seen that the voltage amplification is greater when using a pentode in place of a triode valve under similar conditions of h.t. voltage and load resistance. With a smaller input signal and a larger load resistance this voltage gain can be further improved upon.

Another advantage of the pentode is that the capacitance between the control grid and anode is almost eliminated and as indicated in the section on high frequency operation of transistors in Chapter 12, the presence of capacitance between the input and output circuits can cause a reduction in output.

THE THYRATRON VALVE

The thyratron valve is a triode valve with a small quantity of mercury included in the glass envelope. A test circuit and valve characteristics are shown in *Figure 14.18*. Consider a voltage of E volts applied to the thyratron. The control grid is held negative with respect to the cathode as in the triode valve. Starting with $V_G = -8$ V the valve is non-conducting. V_G is slowly made less negative. In the triode valve this would lead to progressive increases in I_A. With the thyratron nothing happens until a certain critical value of V_G is raeched when the valve suddenly becomes conducting. As with the gas-filled diode, it becomes

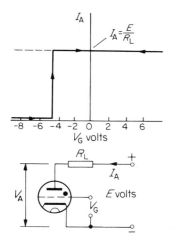

Figure 14.18

virtually a short circuit and sufficient current would flow to destroy the valve if this was not limited by the inclusion of a series connected load resistor.

As the grid voltage is made less negative and then positive there is no change in anode current and more significantly, if the grid voltage is made strongly negative once more this also has no effect. The valve can only be shut off by removing, or reversing the polarity of, the anode voltage. When being fed from a d.c. supply a capacitor current has to be used as in the case of the thyristor (see *Figure 13.12*).

With the valve fed from an a.c. source controlled rectification can be achieved and *Figure 14.19* shows the waveforms of grid and load voltages. The grid is held strongly negative until the valve is required to conduct when it is pulsed to zero or just slightly positive to ensure conduction at low values of anode voltage. *Figure 14.19* shows this happening at point A in one positive half-cycle and at point B in the next. Providing the pulse at point C gives conduction for the whole of the positive half-cycle when the valve is acting as a diode.

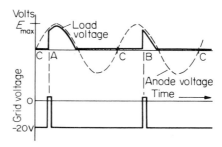

Figure 14.19

The thyratron may be used as a contactor controlled by a small auxiliary switch such as a thermostat or micro-switch operated by a person or process. The auxiliary switch is arranged so that its operation applies or removes the negative grid bias. During the period that the bias is off the valve acts like a gas-filled diode and conducts during each positive-going half-cycle and whilst it is on there is no conduction. The auxiliary switch handles virtually no current at low voltage whilst the valve controls the main circuit power which in a large valve can be in the kilowatt range.

CONTROL OF AN ELECTRON BEAM

Electrostatic

Like electrical charges repel one another while unlike charges attract. An electron being injected between two charged plates as shown in *Figure 14.20* will be attracted towards the positive plate. The force of attraction is proportional to the potential difference between the plates. If the entry velocity of the electron is low it may in fact end up on the positive plate when it will be absorbed as at the anode of a valve. By a suitable choice of entry velocity, plate size and potential difference the electron may be caused to be merely deflected from its original path, and once outside the influence of the electric field between the plates it will travel on in a straight line until further deflected or it impacts with a surface.

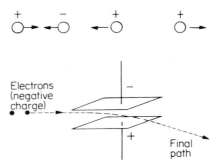

Figure 14.20

Magnetic   A movement of electrons constitutes an electric current and when a current flows in a region in which there is a magnetic field, a force is produced.

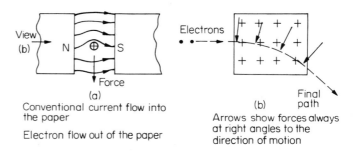

Figure 14.21

Conventional current flow and electron flow are in opposite directions. The force acting on the conductor in *Figure 14.21a* is downwards. This is for a conventional current flow into the paper or electrons flowing out of the paper. In the case of a rigid conductor in which the current comprises countless electrons, the force is uniform along its length in the field and it moves directly downwards mutually at right angles to the field and to the current direction.

    In the case of a single electron, as soon as it enters the magnetic field it suffers a downward force which changes its direction. The force is maintained at right angles to its direction of travel and it follows a circular path as shown in *Figure 14.21b*.

THE CATHODE RAY TUBE

Essentially the cathode-ray tube comprises a conical glass envelope, evacuated to near-perfect vacuum, within which there is:
1. a source of high velocity electrons,
2. a means of focusing the electron beam,
3. a beam-deflection system and
4. a fluorescent screen which glows where it is struck by the beam.

The electron gun   This is very similar to a diode valve with a heated cathode and a flat anode in which there is a hole. Electrons are emitted from the cathode and are accelerated towards the anode which is at a high positive potential. The electrons attain a high velocity and some of them pass through

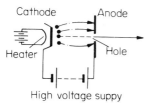

Figure 14.22

the anode hole to travel on until they hit the screen at the end of the tube. Most of the electrons hit the anode and are returned to the cathode.

By adding the grid as shown in *Figure 14.23*, the number of electrons passing through the anode hole can be considerably increased. The grid is a solid metal tube completely open at the cathode end and having a small hole at the end nearest the anode. The grid is held at a negative potential with respect to the cathode and the electrostatic stresses created tend to keep the electrons on a central path as they travel from the cathode. However, making the grid very strongly negative shuts off

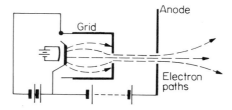

Figure 14.23

the electron flow as in the case of the triode valve so that control of the grid potential affords control of the number of electrons striking the screen and thereby the brilliance of the trace.

The cathode-ray tube with electrostatic focusing and deflection

Figures 14.24 and *14.25* show the arrangement of the electrodes in a practical cathode-ray tube employing electrostatic focusing and deflection. The high-voltage supply to the tube is applied to a train of resistors. The grid/cathode voltage and hence the brilliance of the trace is controlled by adjusting the potentiometer P_1 over a limited range. Electrons are emitted from the cathode and are accelerated towards anode A_1. Electrons which pass through the hole have before them another cylindrical anode A_2 which, acting with anodes A_1 and A_3, produces electrostatic stresses which focus the electrons into a slightly convergent beam as was the case in the grid cylinder. The degree of convergence is adjusted by a limited movement of potentiometer P_2. This adjustment will cause the trace on the screen to be very thin and sharp

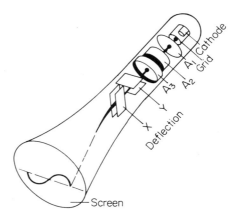

Figure 14.24

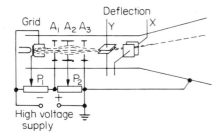

Figure 14.25

or wide and woolly. Deflection of the beam is achieved by using two sets of parallel plates which are fed from two separate supplies. For example the X-plates may be fed from a timebase circuit which causes the beam to sweep across the screen in a horizontal direction whilst an alternating signal is fed to the Y-plates causing vertical displacement. A trace as shown on the tube end in *Figure 14.25* will then be seen.

The electrons strike the screen at high velocity, causing the phosphors with which it is coated to glow, and then return to the supply through the earth connection. Various colours are possible. Television receivers have phosphors which glow blue/white in monochrome sets whilst for colour, three different phosphors are used which glow red, blue and green respectively. Cathode-ray tubes for laboratory and industrial use generally have blue or green traces.

The length of time that the phosphor continues to glow after the electron beam has passed by is important. In television sets this time interval is small since we view 25 different pictures each second, but in the laboratory when transient phenomena are being examined or for radar where the image must persist from one scan until the next, the tube must carry the image long after the beam has passed by. Phosphors have been developed which glow bright yellow and these will display a trace for several seconds. Orange phosphors are also used for some purposes.

In alternative equipments a recording technique is used to give a long-lasting blue/green display.

The cathode ray tube with magnetic focusing and deflection

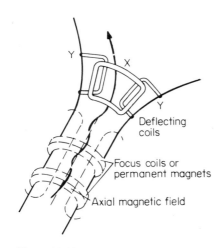

Figure 14.26

Electrons are emitted and accelerated in the same manner as in the electrostatic tube. External coils carrying direct current or permanent magnets are used to produce an axial magnetic field as shown in *Figure 14.26*. Electrons travelling along the axis of the tube, and hence parallel with the lines of magnetic force, are unaffected. Electrons which cross the lines of force at an angle have a component of velocity across them and will suffer a deflecting force as shown in *Figure 14.21*. This imparts a circular motion, which together with the original forward velocity, causes the electron path to be helical. By adjusting the strength and position of the focusing magnets, all such helices may be made to end at the same point on the screen so producing the necessary sharp trace.

The beam is deflected by the magnetic fields produced by currents flowing in two pairs of coils, X and Y, mounted externally at the flare in the tube. Relatively large powers are required to provide the fields, and because of the inductance of these windings the range of operating frequencies must be limited. The main application of this type of tube is for picture display in the television receiver where the frequency of the currents in the Y-coils determines the number of pictures presented per second and that in the X-coils the number of lines per picture.

PROBLEMS FOR SECTION 14

(4) Why does thermionic emission not take place in air?

(5) Explain the mechanism whereby the gas-filled diode conducts more current than the vacuum valve.

(6) How is the emission of electrons maintained in a cold-cathode device once it has been initiated?

(7) What external indication is there that gaseous ionisation has taken place in a valve?

(8) With what solid state device may the triode valve be compared in many respects?

(9) What advantages has the pentode valve over (a) the triode (b) the tetrode?

(10) What are the essential differences between the thyratron and the thyristor especially concerning the grid/gate conditions for
(a) the non-conducting state
(b) the conducting state?

(11) Draw a sketch showing the essential features of a cathode ray tube employing electrostatic focusing and deflection. Describe the principle of operation.

(12) Why is electrostatic deflection used in laboratory cathode ray oscilloscopes whilst magnetic deflection is used in television receivers?

(13) Describe the methods of adjusting the focus on a cathode ray tube when equipped with (a) electrostatic and (b) electromagnetic focusing.

(14) Explain why the persistence of the trace (afterglow) on a cathode ray tube is critical quoting some applications where
(a) a short period and (b) a long period is desirable.

15 Photoelectric devices

Aims: At the end of this section you should be able to:
Identify the wavelengths and bandwidths of visible radiation in the electromagnetic radiation spectrum.
Compare the relative response of different light-sensitive materials.
State the principles, characteristics and applications of photoconductive, photojunction and photovoltaic devices.
State that visible radiation may be emitted from certain forward-biased p–n junctions.
Compare light-emitting diodes with conventional filament lamps.
Describe the applications of photoelectric devices in the detection of hot and cold bodies.
Describe the principle of optical isolation and state typical applications.

VISIBLE RADIATIONS   Electromagnetic radiations with frequencies between 4×10^{14} Hz and 7.5×10^{14} Hz are detected by the human eye and give the sensations of a range of colours from red to violet respectively. Since light travels at a speed of 3×10^8 m/s (approximately) the wavelengths of the radiations, the distance that light travels in the periodic time of one cycle, is found by dividing the speed of propagation by the frequency.

$$\text{Wavelength of red light} = \frac{3 \times 10^8}{4 \times 10^{14}} = 0.75 \times 10^{-6} \text{ m}$$

$$\text{Wavelength of violet light} = \frac{3 \times 10^8}{7.5 \times 10^{14}} = 0.4 \times 10^{-6} \text{ m}$$

Light and illumination are dealt with more fully in Chapter 16.

PHOTOELECTRIC DEVICES   In 1887 Hertz discovered that it was very much easier to cause an electrical discharge to occur between electrodes which were situated in bright light than it was when they were in the dark. This discovery led others to investigate what is known as the photoelectric effect, the connection between light energy and electrical energy.

It was found that light energy caused electrons to become more mobile so that they were able to leave the surface of some substances, while in others voltages could be generated or a change of resistance detected. Similar effects can be produced with heat.

Photoelectric devices are classified as photoemissive, photovoltaic or photoconductive.

PHOTOEMISSIVE CELLS   The photoemissive cell is a diode comprising a coated metallic cathode enclosed in a vacuum tube and which emits electrons when it is illuminated. As in the thermionic diode, the electrons are attracted to an anode which is held at a positive potential and a current flow is set

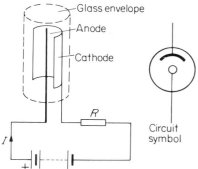

Figure 15.1

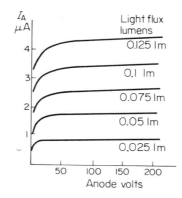

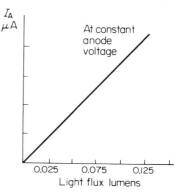

Figure 15.2

up as shown in *Figure 15.1*. The current is proportional to the total light flux at the cathode which therefore has to be large and evenly illuminated to give the optimum output. Typical characteristics are shown in *Figure 15.2*.

Several different cathode coating materials are available each giving its best output when illuminated with light of a different colour. A silver-plated cathode with caesium oxide present has a fairly good response to red light and is used in conjunction with artificial light from incandescent light bulbs. Antimony/caesium cathodes respond well to the blue/violet end of the spectrum and are used in daylight. A response similar to that of the human eye is obtained by adding bismuth to the antimony/caesium cathode.

Using a light filter of a particular colour in front of the tube changes the response in the same way that by using a filter in front of a camera lens the colour values are changed on film.

Adding a small quantity of inert gas to the tube to give a pressure of about 1 mm of mercury causes a much larger current to flow in the external circuit. This is due to ionisation of the gas (see 'The gas-filled diode', Chapter 14). A dot is added to the circuit symbol to indicate gas filling.

Applications

The output from the tube is very small being only a few microamperes and this cannot generally be used to operate a relay directly. Using a d.c. source as shown in *Figure 15.1* a voltage is developed across the load resistor R. This voltage can be used as a bias on the control grid of a triode or pentode valve. Switching the illumination on and off causes the bias value to vary between a relatively large value and near zero so that the valve alternates between the non-conducting and conducting states. A relay connected in the anode circuit will pull on and drop off according to the light conditions.

A light source is situated such that a beam of light is projected on to the cell and the arrangement may be used as a burglar alarm or for counting articles passing a particular point as on a production line. Anything interrupting the beam of light causes the valve current to change and the relay to operate. The relay can ring a bell, operate an automatic telephone calling system or pulse a counting circuit according to the application.

Cinema projection equipment uses the photoemissive cell to convert the varying light output from an optical sound track into sound.

Natural gas burns with a blue flame which gives very little visible light but appreciable ultra-violet. A photoemissive cell with an antimony/caesium cathode can detect ultra-violet radiation and is used to give flame failure warning if for example a pilot light goes out. This would prevent the main valve from opening so releasing large quantities of gas which would not be ignited directly with the risk of an explosion.

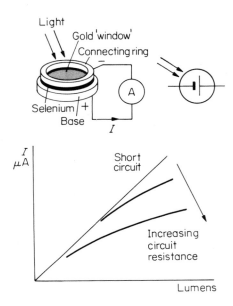

Figure 15.3

PHOTOVOLTAIC CELLS

A selenium cell is shown in *Figure 15.3*. The selenium is deposited on a base of steel or aluminium and then an extremely thin layer of gold added. Light travelling through the gold strikes the gold/selenium junction and produces an electromotive force. The current flowing in the external circuit is a function of the total light flux incident to the surface of the cell. The response matches that of the human eye quite closely.

Figure 15.4 shows the construction of a copper oxide/copper cell which has its greatest sensitivity at the red end of the spectrum.

Silicon and germanium p-n junctions are also light sensitive. There are two possible arrangements, a very thin layer of p-type material on an n-type base or vice versa. These are shown diagramatically in *Figure 15.5*. They are sensitive at the red end of the spectrum.

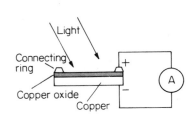

Figure 15.4

Applications

The p-n silicon photovoltaic cells are extremely small and are used to read information from punched cards or paper tape for the control of machine tools, for example. A light source above the punched material shines through the holes on to a matrix of silicon cells. Only those which are illuminated produce e.m.f.s and these voltages provide information used to control the operations. A new card or the paper tape moving on produces a different set of instructions.

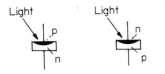

Figure 15.5

Selenium cells with their response similar to that of the eye are used in colour comparators and in commercial light meters in which the output voltage drives a microammeter calibrated in the required units, lux for example.

PHOTOCONDUCTIVE CELLS

The photoconductive cell has the property of variable resistance according to the light intensity (Note: not total luminous flux in this case) at its surface. In the dark it may have a resistance of several megohms while when brightly lit the resistance will fall to a fraction of a megohm. Large changes in resistance are achieved by arranging the electrodes in the form of interleaved combs as shown in *Figure 15.6*. This arrangement gives a large area of active material within a small overall size.

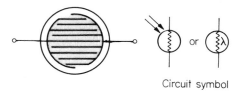

Circuit symbol

Figure 15.6

Materials used include cadmium sulphide, cadmium selenide, indium antimonide, lead sulphide and lead selenide. These cells can carry tens of milliamperes continuously and may be used to control relays directly.

The p–n junction may be used as a photoconductive device by reverse biasing it as shown in *Figure 15.7*. It is then known as a photodiode and the intensity of the light falling on the junction determines the magnitude of the reverse current flow.

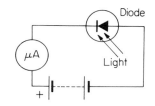

Figure 15.7

Applications

All bodies emit infra-red radiation, the wavelength of the emission being dependent on the temperature of the surface. Photoconductive cells are able to detect small variations in radiation from a general background level. Different materials have their maximum sensitivities at wavelengths corresponding to different temperatures so that by a suitable choice of material extreme sensitivity can be achieved over a very narrow temperature band.

Militarily this is very significant since a device to detect the infra-red radiation from a human face, a glowing cigarette end or gun barrel at a range of hundreds of yards gives very effective night vision.

Civil applications enable hot machine bearings to be detected, hot joints in overhead lines to be located whilst standing on the ground or flying overhead in an aircraft; and the condition of thermal insulation of buildings to be determined by an infra-red scan of the outside.

LIGHT-EMITTING DIODES (LED)

As we have already seen, when a p–n junction is excited with light a voltage is generated and a current will flow in an external circuit. The reverse process is also possible. When a p–n junction is forward biased using an external battery, light may be emitted from the junction. The quantity and colour of the light are dependent on the current flowing and the material from which the junction is made. The process is called photoluminescence.

Junction diodes for this purpose are formed in a similar manner to the photovoltaic selenium cell using gallium arsenide to produce red light, gallium phosphide doped with slight impurities to produce red or green light or a combination of both when yellow light is seen.

The forward voltage drop is in the region of 1.3 V and the light output is determined ultimately by the capability of the diode to dissipate the heat produced by the current flow. The efficiency of conversion is considerably greater than that of the incandescent electric lamp and an LED of a few milliwatts rating can replace a 5 W indicator lamp in equipment such as television sets where they are used to indicate channel selection.

The LED can respond to current variations at over one million times per second so that it can be used wherever a rapidly changing light output in response to an electrical signal of high frequency is required. The production of sound track on film is an example where the sound frequencies will lie between a few hertz and about 20 kilohertz. As a comparison the incandescent light bulb delivers a light output with little flicker at all frequencies above about 30 Hz.

PHOTOTRANSISTORS AND THYRISTORS

The phototransistor has two junctions as in the bipolar junction transistor and is arranged in its capsule so that light can shine on the base/collector junction. When so illuminated the transistor becomes conducting, and it requires no electrical connection to its base. It can respond to changes in light level at up to about 50 000 times per second.

The photothyristor or LASCR (light actuated silicon controlled rectifier) requires a light input to cause it to become conducting instead of the gate pulse used with the normal silicon controlled rectifier.

Both these photo devices may be built into a capsule which contains a small light bulb or light-emitting diode when the capsule has four terminals, two for the light and two for the device. There are also types with a small window to allow for independent illumination.

Either device allows complete electrical isolation of the input signal from the output circuit. The lamp or other light source operates completely independently of the load and there is no danger of feedback or inadvertent connection.

The phototransistor is used in photoelectric counters, alarm circuits, pattern matching and other printing processes, edge detecting in paper making and automatic switching of lighting.

The LASCR is essentially a light-operated relay and switches power on and off in response to a light signal. It may operate as a rectifying device or to control a.c. power as in *Figures 13.8* to *13.10*. *Figure 15.8* shows the circuit symbols.

Phototransistor   LASCR

Figure 15.8

PROBLEMS FOR SECTION 15

(1) The frequency of green light is 5.8×10^{14} Hz. What is its wavelength?

(2) What is the photoelectric effect?

(3) What three effects may be detected in different substances when they are exposed to light?

(4) How does adding a little gas to a photoemissive cell increase its output current?

(5) Under what circuit conditions may certain p–n junctions be used (a) as photoconductive cells, (b) to emit light?

(6) What advantages has an LED over a filament lamp?

(7) How does the circuit resistance affect the output current from a photovoltaic cell?

(8) List some of the applications of:

(a) the photoemissive cell,

(b) the photovoltaic cell,

(c) the photoconductive cell.

(9) What advantages have phototransistors and thyristors over normal types?

16 Illumination

Aims: At the end of this section you should be able to:

*State that light is emitted by hot tungsten wire in an incandescent lamp
and an electrical discharge in a gas.*

Compare efficiencies and colour spectra of various lamps.

Define flux, intensity, illuminance and luminance.

State the inverse square law.

Calculate the illuminance of a surface using $E = I \cos \theta / d^2$ *and a polar
diagram.*

*Calculate the number of luminaires needed to provide uniform lighting
using the lumen method together with maintenance and utilisation
factors.*

Describe the effects of diffusing and prismatic luminaires.

*State the conditions which affect the amount of daylight reaching a
point in a room.*

White light may be resolved into a band of colours called the spectrum
ranging through red, yellow, green, blue and violet by passing it through
a glass prism as shown in *Figure 16.1*. The same effect occurs naturally
when sunlight falls on water droplets, forming a rainbow against a background of dark cloud.

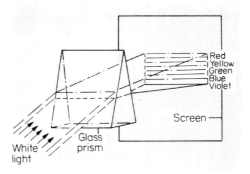

Figure 16.1

These colours are due to electromagnetic radiations similar to radio
waves but of a much higher frequency.

Red light has a frequency of 4×10^{14} Hz

Violet light has a frequency of 7.5×10^{14} Hz

By comparison, radio transmissions have frequencies starting at about
1×10^5 Hz in the long-wave band extending to about 1×10^8 Hz in the
very-high frequency band. Television signals need a higher frequency
still, for example 8×10^8 Hz, and in some ways these act similarly to
light, it being difficult to receive signals over a hill for example.

The speed of all electromagnetic radiations such as light and radio is
3×10^8 m/s (approximately).

At frequencies slightly lower than red there is infra-red radiation which, although invisible, can be detected by the human skin since it is warming. At the other end of the spectrum at frequencies above that of violet light, there is ultra-violet radiation which also has an effect on the human skin, turning it brown as a means of protection. Excessive doses of ultra-violet radiation can be dangerous, burning the skin and damaging the eyes.

The essential source of light and heat for the earth is the sun which provides radiations from ultra-violet, right through the visible spectrum, and into the infra-red region. Much harmful radiation is absorbed by the earth's atmosphere and never reaches the ground, but in areas where this is very clear there are still dangers and eye protection is worn by mountain climbers and skiers for example.

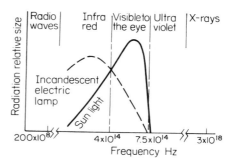

Figure 16.2

Many solid bodies emit light when raised to a sufficiently high temperature. The ordinary incandescent light bulb, which contains a fine tungsten wire being heated by the passage of an electric current, is a good example. The energy supplied is converted into heat and light and *Figure 16.2* shows the various frequencies of radiations in the output from such a bulb. Note the relatively large amount of heat produced and how little of the radiation is in the visible region.

LIGHTING STANDARDS

Since measurements of illumination were first made it has been necessary to have a standard with which the various light sources could be compared.

The original standard was the wax candle. The exact level of illumination was difficult to reproduce each time it was required since the grade of wax, the material for the wick and its burning length had to be specified and accurately controlled. This standard was followed by a lamp burning pentane gas and later by a standard incandescent lamp.

In 1948, with the introduction of SI units, the platinum standard was adopted. It has already been noted that when solid bodies are heated they can emit light. The nature of the light emitted depends on the temperature and this is rather difficult to determine accurately. The problem is overcome by using a material which emits light at a temperature corresponding to that at which it solidifies. It therefore exploits the phenomenon that when materials change state, from liquid to solid or liquid to gas, they do so at constant temperature. A common example is that of water as it freezes into ice or evaporates into steam.

Platinum melts or freezes at 1773°C and at this temperature emits light since it is white hot. The platinum, contained in a small tube within an insulated box, is melted electrically. During the time it is solidifying the temperature remains constant so that the light intensity at a small viewing hole at the top of the box is constant. This light is used as the standard and other sources of light are compared with it. Another reason that platinum is ideal for the standard is that it remains chemically unchanged during successive melts.

LUMINANCE AND LUMINOUS INTENSITY

The luminance of the viewing hole in the standard is defined as 60 candelas per square centimetre (60 cd/cm^2). Let us examine what this means.

Luminance can be interpreted as a measure of the discomfort caused by looking at a light source. Imagine trying to look directly at the noonday sun. It will cause pain and may indeed damage the eye if there is prolonged exposure. The sun has very high luminance, estimated to be about 160 000 cd/cm^2 at noon in summer at Greenwich. In contrast a 1500 mm, 80 W fluorescent tube commonly used for lighting offices and shops has a luminance of about 1 cd/cm^2 and this may be viewed without risk. Imagine now being in a completely darkened room when a small hole is cut in one of the blinds through which the bright sun may be viewed. The amount of light coming into the room through that hole would be minimal and it would be difficult to see things in the room. If however a single fluorescent tube is switched on in the room, far more light is provided than by the sun shining through the small hole.

It can therefore be deduced that light radiation capability or luminous intensity is the product of luminance and area. Thus in the above example the sun with its very high luminance is projecting light through a small area and thereby providing less light than a fluorescent tube of low luminance but with a large surface area.

Since the platinum standard has a luminance defined as 60 cd/cm^2, a hole in the top of the standard having an area of 1/60 cm^2 would result in a light radiating capability of 1 cd. The symbol for light radiating capability or luminous intensity is I so in this case, I = 1 cd.

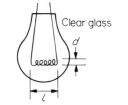

Clear glass

d

Projected filament area $l \times d$

Figure 16.3

Example (1).   The filament of a 100 W clear incandescent lamp has a luminance of 650 cd/cm^2. If the filament has a projected area of 0.1 cm^2 as shown in *Figure 16.3*, what is the light radiating capability in candelas?

I = luminance × area
  = 650 × 0.1
  = 65 cd.

Instead of using clear glass, the interior surface may be treated to create the pearl lamp. This makes the filament appear to have a larger area. If the light radiating capability is unchanged the luminance must have decreased. Where the lamp can be viewed directly, the pearl lamp would be preferred. Alternatively where the lamp is enclosed in some form of shade or diffuser, making direct viewing impossible, the clear lamp would be quite suitable.

THE LUMEN AND THE STERADIAN

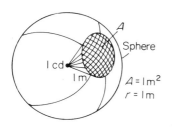

Figure 16.4

Figure 16.4 shows a light source with uniform luminous intensity of 1 candela at the centre of a sphere of radius 1 metre. A uniform source is considered to radiate light equally in all directions but this is not achievable absolutely in practice since all lamps have a dark area where the energy is fed in.

If we consider an area A on the surface of the sphere of 1 square metre, the solid angle so formed is called a steradian, symbol ω.

The amount of light from the source which passes through the cone, i.e. within the solid angle of 1 steradian, is 1 lumen.

The amount of light is called luminous flux and has symbol Φ. Since the radius of the sphere is 1 metre. Its surface area can be calculated using

$$\text{Area} = 4\pi r^2 \text{ metre}^2,$$

$$\text{Area of the sphere} = 4\pi \text{ metre}^2.$$

The total solid angle within the sphere = 4π steradians.

Because there is a luminous flux of 1 lumen in each steradian from a 1 candela source, the total light flux from such a source must be 4π lumens. In *Figure 16.4*, 1 lumen is falling on 1 square metre of area giving a level of illumination or illuminance at the surface of 1 lux. Illuminance has the symbol E. Hence $E = 1$ lux.

In general where the source has luminous intensity I candela, the total luminous flux through ω steradians is $I\omega$ lumens.

> *Example (2).* A lamp has a uniform luminous intensity of 100 cd. What is the total light output of the lamp in lumens?
>
> If all of this light is reflected on to an area of 10 m², what is the illuminance of that area?
>
> Total luminous flux = $100 \times 4\pi$ lumens
>
> $$= 1\,256.7 \text{ lm}.$$
>
> On an area of 10 m² this results in an illuminance of
>
> $$\frac{1\,256.7}{10} \text{ lm m}^2$$
>
> Since the number of lumens falling on 1 m² has been defined as the illuminance in lux,
>
> $$E = 125.7 \text{ lx}.$$

THE INVERSE SQUARE LAW

In *Figure 16.5*, the source of illumination has a luminous intensity of I candela. The level of illumination at surface A_1 is due to I lumens on 1 square metre since the solid angle is 1 steradian.

$$E = I \text{ lux}$$

The light rays are all meant to be at right angles to the surface being illuminated. If the surface A_1 is removed so that all the light rays which illuminated it now travel two metres to illuminate surface A_2, the luminous flux has spread over four square metres. There are now I lumens on 4 square metres.

$$E = \frac{I}{4} \text{ lumens per metre}^2 \text{ or lux.}$$

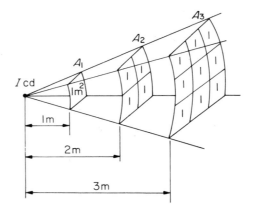

Figure 16.5

For surface A_3 the same luminous flux is distributed over nine square metres.

$$E = \frac{I}{9} \text{ lux.}$$

Generally then, the illuminance $E = I/d^2$ lux where d is the distance of the illuminated surface from the source in metres.

THE COSINE LAW   In the previous section the light rays were all normal to the surface being illuminated. In the case of street lighting for example this would be true only for the position immediately below each lamp.

Consider a single lamp and a point on the road surface some distance from the base of the lamp as shown in *Figure 16.6*. At this range the rays of light are considered to be very nearly parallel. The level of illumination on the area of one square metre shown normal to the light rays is I/d^2 lux. As this area is tilted through the angle $\theta°$ so that it lies in the horizontal plane, the same luminous flux now illuminates an area of $1/\cos\theta$ metre^2. The level of illumination on this increased area now becomes

$$\frac{I}{d_2} \div \frac{1}{\cos\theta} = \frac{I\cos\theta}{d^2} \text{ lux.}$$

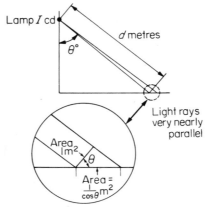

Figure 16.6

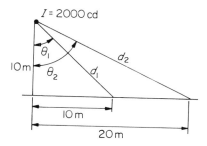

$I = 2000$ cd

10 m

10 m

20 m

d_2

d_1

θ_1

θ_2

Figure 16.7

Example (3). Calculate the level of illumination (illuminance) on a horizontal surface due to a single lamp with uniform light radiating capability of 2000 cd mounted 10 m above the surface at points:
(a) immediately below the lamp,
(b) 10 m from this point on the horizontal surface,
(c) 20 m from the original point on the horizontal surface.

(a) Since the lamp has uniform light radiating capability, the value 2000 cd may be used throughout.

Immediately beneath the lamp the light rays fall normally on the surface.

$d = 10$ m

$$E = \frac{2000}{10^2} = 20 \text{ lux}$$

(b) $\tan \theta_1 = \frac{10}{10} = 1$. Thus $\theta_1 = 45°$. This may be obtained by scale drawing if required.

By Pythagoras' theorem: $d_1^2 = 10^2 + 10^2 = 200$

$$E = \frac{2000}{200} \cos 45°$$

$$= 10 \times 0.707$$

$$= 7.07 \text{ lux.}$$

(c) $\tan \theta_2 = \frac{20}{10} = 2$ Thus $\theta_2 = 63.5°$

$d_2^2 = 10^2 + 20^2 = 500$

$$E = \frac{2000}{500} \cos 63.5° = 1.784 \text{ lux.}$$

Example (4). A light fitting with uniform light radiating capability is mounted 12.5 m above level ground. At a point on the ground 20 m from the base of the lamp standard the illuminance is 2 lx. Determine: (a) the light radiating capability of the lamp in candelas, (b) the total lumen output of the lamp.

THE POLAR DIAGRAM

A lamp without any means of directing the light produced, radiates a good deal of its light in directions other than those required, on to the ceiling for example, or in the case of road lighting, up into the sky. If a reflector or some form of shade or lantern is used in conjunction with the lamp, light can be radiated efficienctly in those directions required.

The light radiation capability of the fitting is now no longer uniform and in order to calculate the illuminance of a surface it will be necessary to use the value of luminous intensity in the direction towards the surface, e.g. along the lines inclined at θ_1 and θ_2 to the vertical in *Figure 16.7*. A polar diagram provides information on the luminous intensities in particular directions.

Polar diagrams for particular lamps and reflectors may be obtained from the lamp manufacturers. Alternatively they can be obtained experimentally by using the following procedure.

A lamp, optionally with reflector, is mounted in a room with matt black walls and in which there is no other source of illumination. A light meter is situated at a known distance from the lamp filament. This distance is often one metre to make the calculations easy. The arrangement is shown in *Figure 16.8*.

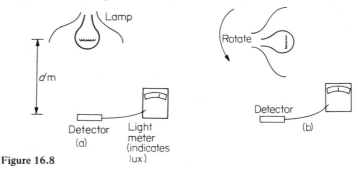

Figure 16.8

The light meter measures illuminance due to the lamp at the light-sensitive surface of the detector.

Illuminance $E = \dfrac{I}{d^2}$ lux

If the distance involved is one metre then the meter reading in lux is numerically equal to the luminous intensity I of the source. In *Figure 16.8a* the luminous intensity in a direction along the axis of the lamp is being found.

If the lamp is turned through $90°$ as shown in *Figure 16.8b* the luminous intensity of the lamp at right angles to its axis is being found. The process is carried out at intervals of $10°$ or $20°$ throughout a complete revolution. Thus the luminous intensity of the lamp and its reflector in any particular direction is evaluated. A plot of the results on polar graph paper is referred to as the polar diagram. Such a curve for the lamp and reflector in *Figure 16.8* is shown in *Figure 16.9*.

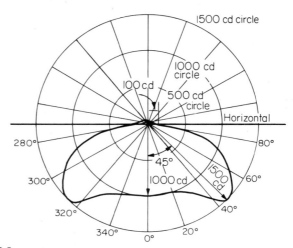

Figure 16.9

The light radiating capability of the fitting is scaled from the origin 0 in the direction required. For instance, directly downwards it is 1000 candelas whilst in the horizontal direction it is only 100 cd. At 45° to the horizontal it is 1500 cd.

The light radiating capability above the horizontal is virtually zero since this particular fitting is designed to project most of the light produced on to the floor.

REFLECTION AND REFRACTION OF LIGHT

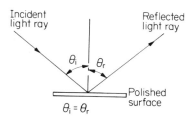

Figure 16.10

Light may be directed to the required area by reflection or refraction. Reflection of light takes place to some extent at most surfaces. If the surface is highly polished, as with a mirror, the reflection is said to be 'specular'. The angle of incidence is equal to the angle of reflection as shown in *Figure 16.10*. Specular reflection is used in the parabolic reflector which can be found in floodlighting fittings or luminaires, in cinema projectors and searchlights etc.

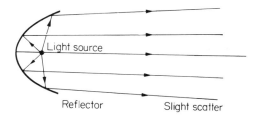

Figure 16.11

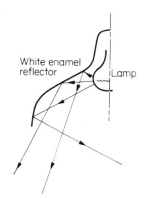

Figure 16.12

The white enamelled steel industrial reflector reflects most of the light downwards as shown in *Figure 16.12*, distributing it over a wide area. Where the surface is matt finished, diffused reflection takes place. This gives a softer lighting effect although it is not so efficient as specular reflection due to light absorption by the surface. The colour of the surface also affects reflection, light colours obviously reflecting more light than dark ones.

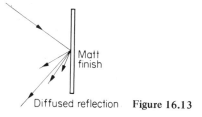

Figure 16.13

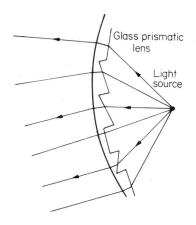

Figure 16.14

Refraction or bending of light rays takes place when they pass through a prism as shown in *Figure 16.1*. The refracting lantern or luminaire is used for street lighting and in motor vehicle headlamps. An array of prisms may be seen on the inside of the glass as shown in cross-section in *Figure 16.14*. By a combination of these methods a polar diagram of almost any shape can be produced enabling light to be projected in the direction where it is most needed. If the object of a luminaire is to diffuse the light so as to reduce the luminance of the source, then diffusing glass globes may be used. These will be finished pearl or opal on the inside or again may make use of much smaller prisms which have the effect of scattering the light as shown in *Figure 16.15b*.

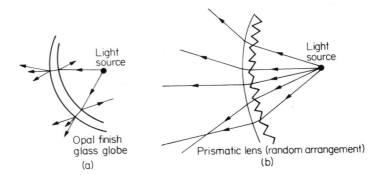

Figure 16.15

USE OF THE POLAR DIAGRAM

Suppose the surface level of illumination or illuminance provided by a lamp in a specific type of luminaire is required. The polar curve is plotted or obtained from the manufacturer. A typical example is shown in *Figure 16.16*.

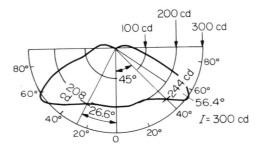

Figure 16.16

The following example illustrates the method to be adopted.

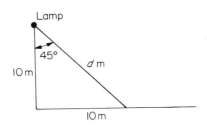

Figure 16.17

Example (5).   A single lamp in a reflector has a polar diagram as shown in *Figure 16.16*. It is mounted on a lamp standard at a height of 10 m above a horizontal flat plane. Calculate the illuminance on the plane:
(a) immediately below the lamp,
(b) 10 m from the base of the lamp standard.
The first step is to draw the arrangement of lamp, standard, and surface to be illuminated as shown in *Figure 16.17*. The lamp is considered to be situated at the origin of the polar diagram. The light illuminating the spot immediately below the lamp is shining directly downwards through zero degrees on the polar diagram. By direct reading from the curve, the luminous intensity in this direction is 200 cd.

(a) $E = \dfrac{I}{d^2} \cos \theta$ lux and in this case $\theta = 0°$.

$$E = \frac{200}{10^2} = 2 \text{ lux}.$$

(b) At a distance of 10 m from the base of the standard, the light is projected at an angle of 45° from the vertical

$d^2 = 10^2 + 10^2 = 200.$

From the polar diagram, at 45° to the vertical, the luminous intensity of the source is 244 cd.

$E = \dfrac{244}{200} \cos 45°$ according to the cosine rule.

$E = 0.863$ lux.

Example (6). Consider the case of two lamps 20 m apart as shown in *Figure 16.18*.

The mounting heights and polar diagram for each lamp are as in the previous example. Calculate the level of illumination on a line joining the bases of the two standards at a distance of 15 m from one of them. This is point P in *Figure 16.18*.

The illumination at point P comes from two sources. Consider them separately.

From lamp A: $E = \dfrac{300}{10^2 + 15^2} \cos 56.4°$

$\qquad = 0.51$ lux.

From lamp B: $E = \dfrac{208}{10^2 + 5^2} \cos 26.6°$

$\qquad = 1.48$ lux.

Total illuminance due to both sources $= 1.48 + 0.51$

$\qquad\qquad\qquad\qquad\qquad = 1.99$ lux.

Example (7). Calculate the illuminance due to both lamps directly under one of the lamps in Example 6. The polar diagram for each lamp is again as in *Figure 16.16*.

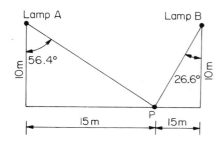

Figure 16.18

INDIRECT ILLUMINATION OF A SURFACE

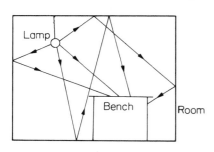

Figure 16.19

The previous examples using the polar diagram show how the illumination at a point may be calculated when that illumination is due to a single source, with no reflecting surfaces nearby, e.g. as in road lighting. If there are two or more sources, again with no reflecting surfaces nearby, then the illuminance present is the sum of the values due to individual sources. With indoor illumination however, this method cannot be used since the distances are small and much of the illumination is the result of reflection from ceilings and walls.

Consider a workshop which requires a given illuminance at the bench tops. The benches receive light in any number of ways as shown in *Figure 16.19*. Some comes directly from the fitting and some by reflection from the walls, ceiling and floor. Note that light can be reflected several times, losing a little by absorption on each occasion.

The amount of light received after reflection depends on the colour and cleanliness of the surfaces involved. It also varies according to the height of the ceiling and the shape of the room. The arrangement of the reflecting surfaces, their condition and colour is taken into account by the use of a 'coefficient of utilisation'.

The coefficient of utilisation is defined as

$$\frac{\text{Luminous flux arriving at the working surface}}{\text{Total luminous flux supplied by the lighting fittings}}$$

This has been determined experimentally by the Illuminating Engineering Society for many different combinations of colour and room proportions and may be found from reference tables in the I.E.S. code. The I.E.S. also specify the requirement for illuminance at the working surface for various types of work and the following table quotes some of these.

Workshops and factories	*Illumination recommended in lux*
Average for general work	100–150
Assembly—large work	100
Radio assembly	200
Engraving	500
Warehouses	50
Drawing offices	300

Example (8).   A workshop 30 m × 40 m is to be illuminated using gas-filled tungsten filament lamps in standard dispersive reflectors. They are situated 2.5 m above the working plane on which an illuminance of 100 lux is required.
The coefficient of utilisation is 0.6.
Calculate the total luminous flux required from the fittings.

100 lux = 100 lumens on each square metre.
Area to be illuminated = 30 × 40 m^2
$$= 1\ 200\ \text{m}^2$$
Luminous flux at the working surface = 1 200 × 100
$$= 120\ 000\ \text{lm}.$$

Since the coefficient of utilisation =

$$\frac{\text{Luminous flux at the working surface}}{\text{Total input of luminous flux}}$$

$$0.6 = \frac{120\ 000}{\text{Input}}$$

Hence, input $= \dfrac{120\ 000}{0.6}$

$$= 200\ 000\ \text{lm}.$$

During the life of an electric lamp its light output decreases due to ageing of the filament or of the phosphors employed in fluorescent tubes. Lamp reflectors also become less efficient due to an accumulation of dust. Inevitably some lamps fail altogether. In a factory it is not usually economic to replace single lamps especially if they are rather inaccessible. It is better to wait until either 5 per cent of the installed lamps have gone out or all the lamps have been in operation for their design life before replacing all the lamps at the same time. This can usually be done when the factory is not in production.

Where a minimum level of illumination is specified, allowance is made for deterioration of lamps by providing a greater level of illumination than this minimum when the installation is new. At the end of the design life of the lamps the output will still be maintained at or above the minimum requirement.

A depreciation factor is used to calculate the required original input. If the illuminance in the workshop of the previous Example 8 must never fall below 100 lux and a depreciation factor of 1.3 is recommended, then the input when new must be 1.3 × 200 000 lumens.

Alternatively a maintenance factor may be used.

$$\text{Maintenance factor} = \frac{\text{light output at the end of the design life}}{\text{light output when new}}$$

Again using data from the previous example, the maintenance factor would have been 0.77.

$$0.77 = \frac{200\,000}{\text{light output when new}}$$

$$\text{Light output when new} = \frac{200\,000}{0.77} = 260\,000 \text{ lumens } (= 1.3 \times 200\,000)$$

$$\text{Depreciation factor} = \frac{1}{\text{Maintenance factor}}$$

Clearly, whenever the maintenance or depreciation factor is used in calculations it must give an original light input which exceeds that at the end of the design life.

It is now necessary to examine the numbers and arrangement of lamps by which the required illuminance may be provided.

Data for gas filled tungsten filament lamps is as follows:

Power, watts	60	100	200	300	500
Total output, lumens	576	1 135	2 600	4 140	7 500

Since a 60 W lamp gives 576 lumens, to provide 260 000 lumens would require 260 000/576 = 452 lamps.

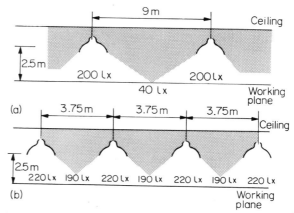

Figure 16.20

Alternatively, using 500 W lamps, 260 000/7500 = 35 lamps would be required.

The effect of having a few large lamps widely spaced is to leave areas of low illuminance between the lamps. This may be seen in *Figure 16.20a*. Generally it is found that spacing the lamps between 1.5 and 2 times their mounting height above the working plane produces an acceptable degree of variation. This is shown in *Figure 16.20b*. Using lamps closer together than this increases the installation costs. Two suitable arrangements of lamps are shown in *Figure 16.21*.

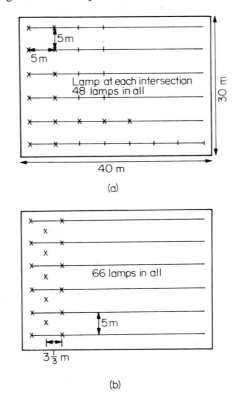

Figure 16.21

The arrangement in *Figure 16.21a* uses 48 lamps each with 500 W rating. The total electrical loading = 48 × 500 = 24 000 W and the luminous flux provided = 48 × 7500 = 360 000 lumens.

The arrangement in *Figure 16.21b* uses 66 lamps each with 300 W rating. The total electrical loading = 66 × 300 = 19 800 W whilst the total luminous flux = 66 × 4140 = 273 200 lumens.

Note that it is not generally possible to satisfy the requirements exactly without having an asymmetrical arrangement of lamps and this would have an unsatisfactory appearance.

TYPES OF LAMP

The incandescent lamp   These lamps have a tungsten filament heated within a glass bulb which has been either exhausted to near-perfect vacuum or contains a small amount of inert gas. It radiates energy over a continuous spectrum as shown in *Figure 16.2*. The proportion radiated as light increases with filament temperature but at best this is only a small fraction of the available energy.

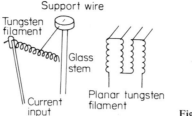

Figure 16.22

Figure 16.22 shows two possible filament arrangements, one for general service lamps and the other for film projectors. Outputs vary from about 9 lumens per watt for the 40 W size to about 14 lumens per watt for the 500 W size.

The high-pressure mercury discharge lamp

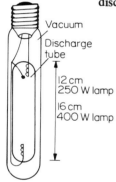

Figure 16.23

In this lamp, current is caused to flow through a small tube in which there is a little inert gas and a few droplets of metallic mercury. The current flows at first in the gas producing enough heat to vaporise the mercury. The ionisation of the mercury vapour produces light (see Chapter 14 for ionisation of gases). The pressure when hot is about 1 bar.

The space between the inner discharge tube and the outer bulb is partially evacuated to minimise the heat loss from the lamp. This is necessary to keep the mercury in vapour form.

The colour of the light is blue/green with no additional radiation corresponding to red and yellow present. However, there is ultra-violet present and if the inside of the outer bulb is coated with a suitable phosphor, the ultra-violet emission causes this to emit a red component so making the light emitted more nearly equivalent to daylight. The uncompensated spectrum is shown in *Figure 16.26*. The light output is about 40 lumens per watt.

The low-pressure mercury (fluorescent) tube

This lamp also contains mercury but runs at a much lower pressure than the type just described. It is also larger physically, being typically 1 500 mm long at 65 or 80 W rating. It runs much cooler at around 40°C. The discharge from the mercury under these conditions contains a substantial amount of ultra-violet. The inside of the tube is coated with phosphors which emit visible radiations (light) as a result of being excited by the ultra-violet radiations from the gas. The colour can be altered by the selection of phosphors, a technique employed in the manufacture of colour television tubes. The light output of the tube varies between 40 and 60 lumens per watt according to colour, 'blue' (north light) tubes giving less light than 'warm white' tubes for the same rating.

The low-pressure sodium discharge lamp

This lamp consists of a U-tube containing a little sodium metal and neon gas at very low pressure. When the supply is switched on, the neon gas conducts current and a bright red light is emitted. Slowly the sodium boils off into the near vacuum of the tube and the ionised sodium vapour emits bright yellow light. The spectrum is shown in *Figure 16.26*. As with the high-pressure mercury lamp, heat must be conserved and therefore the U-tube is supported in a second vacuum

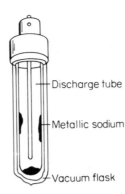

Figure 16.24 Low pressure sodium lamp

flask. If this fractures, the light reverts to the red colour as the sodium solidifies. The light output is about 130 lumens per watt. This lamp is used for road lighting where the colour of the light is not too significant.

The halogen lamp

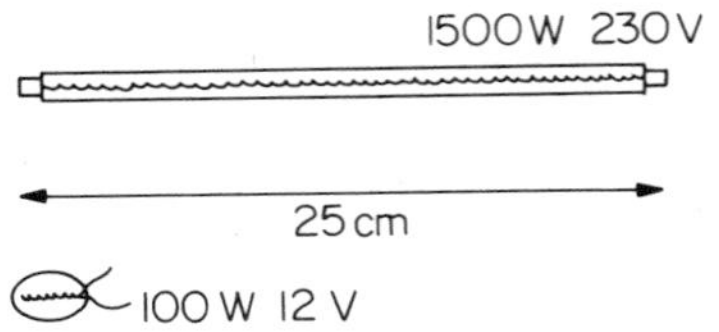

1500W 230V

25 cm

100 W 12 V

Figure 16.25 Tungsten halogen lamps

The conventional tungsten filament decreases in cross-sectional area by evaporation. The tungsten condenses on the bulb wall so reducing the light output. The glass bulb is made large so that the thickness of the deposit is small and the consequent loss of light is minimised. If a little iodine is added to the gas in the bulb, any tungsten which boils off combines chemically with the iodine and may subsequently be returned to the filament. There is no blackening of the bulb and this can be made much smaller than that of a conventional lamp of equivalent rating. Because of their reduced size, these lamps can be placed at the

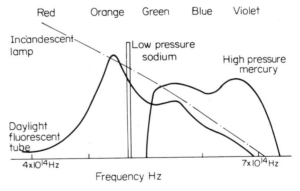

Figure 16.26

centre of focus of a parabolic reflector and they are found in motor vehicle headlamps and floodlights for public areas. The light output is between 15 and 25 lumens per watt.

The high-pressure sodium lamp

If the pressure in the sodium lamp is increased and the temperature raised to about 1 000°C, the discharge becomes much whiter and blends well with light from normal incandescent lamps. This development could not take place until a new material for the tube had been developed. The low-pressure lamp uses a glass tube while this lamp uses an aluminium oxide (alumina) tube which is capable of withstanding the very high temperature. It is used extensively for floodlighting where the warm, slightly pink colour is very attractive. The light output is about 100 lumens per watt.

TYPES OF LUMINAIRE

A selection of luminaires available is shown in *Figure 16.27*.

Types (a) to (d) are used indoors, types (e) to (i) outdoors. They can be diffusing or concentrating and are chosen for these functions, at the same time considering their appearance and the environment in which they are to be used. Outside luminaires must be weather-proof and as

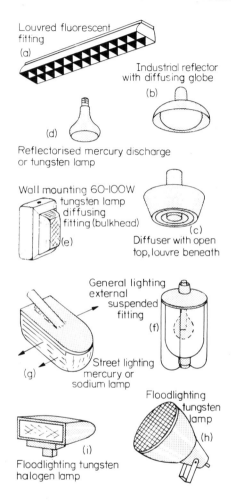

Figure 16.27

far as possible, vandal-proof. The transparent parts are made of pre-stressed glass or impact resistance plastic. To minimise maintenance, the surfaces should be self-cleaning by the action of wind and rain. Therefore any patterns or prisms are on the inner surface of the glass. Electrical connections to them are generally by means of conduit or mineral insulated cables.

DAYLIGHTING

In most buildings use is made of daylight when available, supplemented by artificial light. The correct balance between the two is difficult especially when the rooms are deep when the variation in daylighting from a position near a window to that several metres from a window may be considerable.

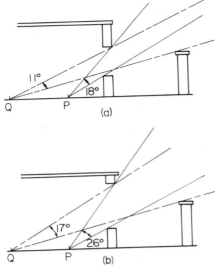

Figure 16.28

The amount of daylight reaching a point inside a building through a window directly from the sky will depend on the area of the sky visible from that point, on the luminance of the sky and the angle at which it arrives on the surface to be lit. As already stated:

$$E = \frac{I}{d^2} \cos \theta° \text{ lux.}$$

Figure 16.28 shows how the arc of sky visible is dependent on the distance inside the room. In *Figure 16.28a* at point P there is 18° of sky visible whilst at point Q there is only 11°. Note how the wall some distance from the building restricts the light input to point Q. If this wall were not there the arc would extend downwards to the first obstruction which might be the window sill or the horizon. In *Figure 16.28b* the effect of a taller window can be seen. More arc of sky is visible from both points. Note also in both cases the difference in the angle of incident light between points P and Q.

Figure 16.29 shows a plan view of the room. Note how the arc of sky visible is limited in the horizontal plane also. At points off the centre line the arc is further reduced.

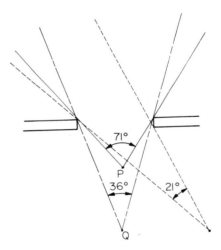

Figure 16.29

The luminance of the sky depends on the season of the year, the hour of the day and the distance of the room from the equator, i.e. on its latitude, since the height of the sun above the horizon depends on these factors. The average level of illumination on a horizontal surface in the open, which is a measure of the luminance of the sky, measured at Greenwich using north light only varies from about 450 lux at 9 a.m. to 1700 lux at midday in January compared with 4150 lux and 5700 lux at the same times in June. In addition, the luminance of the sky will depend on whether it is overcast or not. This may be due to clouds or pollution.

One further factor affecting the amount of light entering a room is the direction in which it is facing, north-facing rooms receiving no direct sunlight. Finally there is a component of light derived by reflection from external surfaces and this depends on the state and colour of those surfaces. This is illustrated in *Figure 16.30*.

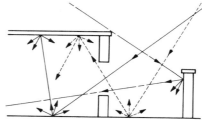

— Light entering the room directly from the sky

-·-Light entering the room by reflection from
an external vertical surface

--- Light entering the room by reflection from
an external horizontal surface

Figure 16.30

Illuminance on a horizontal plane 1 m
above the floor due to north light
through a 1½m wide 3m high clear
glazed window

Figure 16.31

Once light has entered a room by whatever means, the final distribution will depend on the state and colour of the decorations, diffused reflection taking place on most surfaces. *Figure 16.31* shows a typical illuminance curve for a room lit by north light.

PROBLEMS FOR SECTION 16

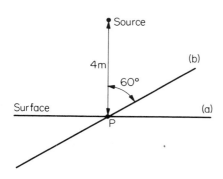

Figure 16.32

(9) The light source shown in *Figure 16.32* has a uniform luminous intensity of 64 cd. The point P on the surface to be illuminated is 4 m from the source. Calculate the illuminance at point P when the surface is in position (a) and then in position (b).

(10) Why are incandescent light bulbs often made from obscured (pearl) glass?

(11) Why is the tungsten planar filament suitable for slide and cinema projectors?

(12) Why does it take approximately 10 minutes for the high-pressure mercury vapour lamp to reach its full brilliance?

(13) What would happen to a low-pressure sodium discharge lamp if its vacuum envelope were removed?

(14) Why is it that the high pressure sodium lamp has been available only in the last few years?

(15) What advantages has the tungsten halide lamp in the field of public lighting and automobile engineering?

(16) What special features should a luminaire have if it is to be used outdoors?

(17) List the ways in which daylight reaches a table top in a room with its only window in the vertical plane.

(18) Upon what factors does the illuminance at a surface in a room depend when lit by daylight only?

(19) A roadway is illuminated by means of lanterns which have luminous intensities in the vertical plane as stated in the table. The standards are 5 m high and 40 m apart.

Angle from the vertical Degrees	Luminous intensity Candela
0	300
10	310
20	330
30	360
40	400
50	450
60	530
65	590
70	685
75	810
80	660
85	0

Calculate the illuminance due to the nearest two luminaires at a point on a line on the ground joining two standards, 10 m from one of them.

(20) A room 20 m × 40 m is equipped with 50 fluorescent tubes each giving a total output of 3400 lumens when new. The coefficient of utilisation for the room is 0.56 and a depreciation factor of 1.25 has been allowed. Calculate the level of surface illuminance (a) when all new tubes have just been installed and (b) just before these lamps are due for complete replacement.

Solutions to problems

SECTION 1   (1) (i) 285.7 A, 16.33 kW; (ii) 28.57 A, 163 W

SECTION 2   (4) 19.05 kV
    (5) 216.5 V
  (17) 24 A
  (20) 0.972 p.u.
  (21) AB, 23.8 A, BC 3.81 A, CD 26.2 A, Min. P.D. at C = 218.43 V
  (22) 50 A load power = 11.794 kW
  (23) AB 89.6 A, BC 69.6 A, CD 39.6 A, DE 29.6 A, EF 20.4 A,
       FG 40.4 A, GA 65.4 A. Min. P.D. at E = 235.9 V.

SECTION 4   (2) 2.68 p/kWh, load factor 0.076
    (4) M.D. = 228.3 kW Average 2.0196 p/kWh
    (6) £2 222.22   £1422 after paying for correction equipment.
    (7) (a) (i) 0.183 (ii) 0.363   (b) 0.0476   (c) 2.384 p/kWh £29.80
    (8) 2.14 p/kWh
    (9) 400 kW at £14 = £5600 p.a.
  (11) (a) £25 756.3
       (b) M.D. saving £1 821.22 Cost £1042 Overall saving £778.8
       (c) M.D. saving £473.48 Cost £737 Not sound.

SECTION 5   (2) Factor for 6 circuits 0.55 1 circuit 54.5 A 3 circuits 37.6 A
    (4) 2.5 mm² 10 mm² 4 mm²
    (9) 16 mV/A/m, 0.24 V, 3.6 W
  (10) 10.35 A
  (11) 10 kW, 1000 newtons per metre run.
  (15) 0.875 A
  (16) 71.4 A/mm² 18.87 A/mm²

SECTION 6   (3) (a) 0.83 A (b) 332 W (c) 0.638 (d) 150 turns (e) 0.003 Wb.
    (4) 40 A
    (5) (a) 1.03 A (b) 2.675 A (c) 3.5 A (d) 43.7°
    (8) (a) 0.97
       (b) copper loss = 180 W efficiency 0.98
       (c) Copper loss = 101.25 W efficiency 0.974
  (12) 80
  (14) (a) 112.5 V (b) 2.778 A (c) 187.5 W.
  (17) 100 W
  (18) (a) 98 W (b) 200 W
  (19) 100 W
  (20) 0.985
  (23) (a) 0.67 (b) 0.3
  (26) (a) 208.33 A (b) 4.545 A (c) 0.673 (d) 5.6 A

SECTION 7   (3) (a) 180 V (b) 360 V Power = 14 400 W in both cases.
              (8) (a) 504 V (b) 29.14 rev/s
              (9) (a) 1008 V (b) 14.57 rev/s
             (14) (a) 129 V (b) 387 V
             (15) 26 rev/s (E = 510 V, kΦ = 19.16)
             (16) (a) 10.83 rev/s (b) 0.036 Wb.
             (19) (a) Add 9.7 Ω, (b) Add 4.7 Ω.
             (21) Output = 19 150 W efficiency = 0.837 p.u.
             (22) Input = 25 832 W efficiency = 0.852

SECTION 8   (5) 5×10^{-4} Ω.
              (6) 999 995 Ω.
             (12) 424 W

SECTION 10   (2) (a) 910 000 1%   (b) 390 000 4%
                  (c) 10 20%   (d) 4700 5%   (e) 0.47 2%
              (3) (a) Red, Green, Black   (b) Green, Violet, Black
                  (c) Brown, Green, Brown   (d) Green, Grey, Orange.
              (4) ± 0.001 $\Omega/\Omega/$°C, 1.05 to 0.95 Ω. NiCh. + 0.00007$\Omega/\Omega/$°C,
                  1.0035 Ω.
              (6) (a) Brown, Green, Brown, Yellow
                  (b) Green, Grey, Black, Red
                  (c) Green, Grey, Silver, Silver
              (7) (a) 815 Ω 5% (b) 732 Ω 10% (c) 146 Ω 3%
             (11) At 10 μF correct current = 0.314 A. The capacitor is virtually
                  open circuited.
             (12) Z = 5 Ω. X_L = 4.58 Ω L = 14.58 mH
             (13) (a) Thyristor. Terminal 1 = anode, 2 = cathode, 3 = gate
                  (b) JFET. Terminal 1 = gate, 2 and 3 = source and drain
                  (c) pnp junction transistor. Terminal 1 = emitter, 2 = base,
                    3 = collector.

SECTION 11   (3) 5 μF 100 μF
              (4) ΔV = 10 V 3.32% ripple
              (7) I_L = 1.85 A Volt drop = 9.25 V Input = 46.25 V
              (8) (a) 8.99 V (b) 17.98 V
              (9) 9 V r.m.s. = 12.73 V peak (= ripple voltage) ripple frequency = 100
             (10) 126 V 114 V
             (11) (a) 250 μF (b) 31.25 μF
             (12) 0.52%
             (14) (a) 12.5 V 0.1 A 1.2 W
                  (b) 18 V 0.25 A 3 W

SECTION 12   (2) For 80 μA bias. I_C = 1.36 mA r.m.s. I_{gain} = 48.1
                  V_{CE} = 1.096 V r.m.s. V_{gain} = 70.5
                  Power gain = 3.389 For other values of bias, slightly different
                  results
              (4) For Q on the −0.8 V characteristic:
                  V_{DS} = 1.06 V r.m.s. V_{gain} = 7.5 I_D = 0.884 mA r.m.s.
                  Load power = 0.93 mW.

(5) (a) 5 mA (b) 4.95 mA
(6) (a) 5.083 mA (b) 83.3 μA
(12) (a) 89.97 (b) 47 (c) 4 228.6
(13) 916.7 Ω
(15) (a) 35.35   (b) 1.59 mA r.m.s.
       (c) 1.273 V (d) 44.98
       (e) 36   (f) 1 619

SECTION 14   (2) (a) 16.75 (b) 13.75
       (3) Bias -4 V, gain = 12

SECTION 16   (4) 2 948 cd   37 047 lm

(7) $\dfrac{250}{500} \cos 63.4 + 2 = 2.224$ lx

(9) (a) 4 lx (b) 3.46 lx

(19) From the nearer lamp, $E = \dfrac{590}{5^2 + 10^2} \cos 65.3 = 1.97$ lx

From the further lamp, $E = \dfrac{600}{5^2 + 30^2} \cos 80.6 = 0.106$ lx

Total illuminance = 2.076 lx.

(20) New light flux = 95 200 lm $E = 119$ lx At the end, $E = 95.2$ lx.